T0073856

Advance Praise for *Practical Neurophysics*

"Medical professionals and lay people alike will enjoy learning about the history of and physics behind technologies such as electroencephalography, ultrasound, and SPECT imaging. As an engineer working in medical research, it is probably no surprise that I appreciated learning more about this broad range of assessment and treatment technologies but the information is provided in such a way as to be accessible to everyone, including non-engineers and non-physicists."

—Catherine G. Ambrose, PhD (Mechanical Engineering),
Professor, Department of Orthopaedic Surgery,
The University of Texas Health Science Center at Houston

"Non-invasive imaging and electronic instruments are ubiquitous in the practice of neurology and study of neurosciences. This text is an excellent introduction to the principles of operation and underlying physics of those technologies, and is essential reading for those entering the neurosciences or looking to demystify the tools of the field."

—Michael M. Thornton, MS (Electrical Engineering), Endra Life
Sciences; Member, American Association of Physicists in Medicine

"*Practical Neurophysics: The Physics and Engineering of Neurology* is an exceptional guide to understanding the basics of electrodiagnostic studies. This is a very well-written, easy-to-read, one-of-a-kind book for all-level neurodiagnostic professionals including attendings, epilepsy/clinical neurophysiology fellows, neurology residents, and END technologists. This book provides all the necessary basic science applications and knowledge to build a well-functioning neurodiagnostic laboratory."

—Hasan H. Sonmezturk, MD, Neurologist/Epileptologist,
Ascension Sacred Heart Hospital Epilepsy Center

Practical Neurophysics

The Physics and Engineering of Neurology

EDITED BY

EVAN M. JOHNSON

INSTRUCTOR IN NEUROLOGY
VANDERBILT UNIVERSITY MEDICAL CENTER

KARL E. MISULIS

PROFESSOR OF CLINICAL NEUROLOGY AND
CLINICAL BIOMEDICAL INFORMATICS
VANDERBILT UNIVERSITY SCHOOL OF MEDICINE

OXFORD
UNIVERSITY PRESS

Oxford University Press is a department of the University of Oxford. It furthers
the University's objective of excellence in research, scholarship, and education
by publishing worldwide. Oxford is a registered trade mark of Oxford University
Press in the UK and certain other countries.

Published in the United States of America by Oxford University Press
198 Madison Avenue, New York, NY 10016, United States of America.

Library of Congress Cataloging-in-Publication Data
Names: Johnson, Evan M., author. | Misulis, Karl E., author.
Title: Practical neurophysics : the physics and engineering of neurology /
Evan M. Johnson and Karl E. Misulis.
Description: New York, NY : Oxford University Press, [2023] |
Includes bibliographical references and index.
Identifiers: LCCN 2022018769 (print) | LCCN 2022018770 (ebook) |
ISBN 9780197578148 (hardback) | ISBN 9780197578162 (epub) | ISBN 9780197578179
Subjects: MESH: Nervous System Physiological Phenomena | Biophysical
Phenomena | Bioengineering | Nervous System Diseases
Classification: LCC QP361 (print) | LCC QP361 (ebook) | NLM WL 102 |
DDC 612.8/043—dc23/eng/20220701
LC record available at https://lccn.loc.gov/2022018769
LC ebook record available at https://lccn.loc.gov/2022018770

DOI: 10.1093/med/9780197578148.001.0001

9 8 7 6 5 4 3 2 1

Printed by Integrated Books International, United States of America

Preface xi
Contributors xiii
Abbreviations xv

Physics and physiology have the same Greek root and are so closely related, yet they seem distant in the minds of many. Neurology is heavily dependent on physics for the understanding of the functions of the body as well as functions of all of the wonderful technology that we use.

Over the years, we have developed a black-box approach to some of our technology, knowing the inputs and outputs but being agnostic to the throughput. Having a deeper knowledge base of the physics of neurology may not make one a better clinician, but it may give a better appreciation of the science that we embrace.

In this book, we begin with a discussion of foundational physics and physiology, including basic physics principles, then basic electronics. We introduce computers and computer science as they pertain to medical practice. Then we present a more detailed discussion of select technologies with which we engage every day. We conclude Part 1 with a biophysical overview of human neurophysiology.

The second and largest part of this book concentrates on the physics of neurologic technologies, including electroencephalography, electromyography, evoked potentials, ultrasound, computed tomography, magnetic resonance imaging, and select laboratory studies. We close with a discussion of the physics of some select neurologic therapeutics.

We thank our colleagues and our mentors who helped us on this journey. It is our hope that we and our successors will never lose our interest in and understanding of our science.

Evan M. Johnson
Karl E. Misulis

CONTRIBUTORS

Eamon Doyle, PhD
MR Physicist and Assistant
    Professor of Clinical Radiology
Children's Hospital of Los Angeles
Los Angeles, CA, USA

Evan M. Johnson, MD, MS
Instructor in Neurology
Vanderbilt University Medical
    Center
Nashville, TN, USA

Karl E. Misulis, MD, PhD
Professor of Clinical Neurology
    and Clinical Biomedical
    Informatics
Vanderbilt University School of
    Medicine
Nashville, TN, USA

AC	alternating current
ACh	acetylcholine
AD	Alzheimer disease
ADC	analog-to-digital converter
ADEM	acute disseminated encephalomyelitis
AEP	auditory evoked potential
AIDP	acute inflammatory demyelinating polyneuropathy
CAA	cerebral amyloid angiopathy
CDS	clinical decision support
CIDP	chronic inflammatory demyelinating polyneuropathy
CSF	cerebrospinal fluid
CT	computed tomography
CTA	CT angiography
CTV	CT venography
CVS	cerebral vasospasm
DBS	deep brain stimulation
DC	direct current
DNA	deoxyribonucleic acid
DWI	diffusion-weighted imaging
ECG	electrocardiology
ECT	electroconvulsive therapy
EEG	electroencephalography
ELISA	enzyme-linked immunosorbent assay

EMG	electromyography
EMIT	enzyme multiplied immunoassay technology
EVT	endovascular therapy (for stroke)
FDG	fluorodeoxyglucose
FLAIR	fluid-attenuated inversion recovery (MRI sequence)
GBS	Guillain–Barré syndrome
GI	gastrointestinal
GPI	globus pallidus internus
HIV	human immunodeficiency virus
HSV	herpes simplex virus
I	current
ICP	intracranial pressure
IgG	immunoglobulin G
IgM	immunoglobulin M
IPG	implanted pulse generator
LEMS	Lambert–Eaton myasthenic syndrome
LINAC	linear accelerator
MAP	mean arterial pressure
MG	myasthenia gravis
MIBI	sestamibi scan
MRI	magnetic resonance imaging
MS	multiple sclerosis
NCS	nerve conduction study
NCV	nerve conduction velocity
NMJ	neuromuscular junction
NMR	nuclear magnetic resonance
OCD	obsessive–compulsive disorder
PCR	polymerase chain reaction
PD	Parkinson disease
PT	prothrombin time
PW	pulse width
QEEG	quantitative electroencephalography
R	resistance
RCVS	reversible cerebral vasoconstriction syndrome

RF	radio frequency
RNS	responsive neurostimulation
SAH	subarachnoid hemorrhage
SEP	somatosensory evoked potential
SNAP	sensory nerve action potential
SPECT	single-photon emission computed tomography
SRS	stereotactic radiosurgery
TEED	total electrical energy delivered
TMS	transcranial magnetic stimulation
TPA	tissue plasminogen activator (for stroke)
V	voltage
VEP	visual evoked potential
VIM	ventral intermediate nucleus
VNS	vagal nerve stimulation
VST	venous sinus thrombosis
Z	impedance

PART 1

Basic Physics for Neurology

Basic Physics

EVAN M. JOHNSON AND KARL E. MISULIS ■

Atomic physics is the perfect place to start a book on neurotechnology. It all begins with physics. Throughout this chapter and in the rest of the book, we concentrate on the interesting and essential details, without getting too far into the formulaic weeds. We presume a foundation in basic physics, physiology, and some knowledge of neuroscience. In assembling this book, we have brought together physicians, physiologists, and physicists, so we anticipate the readership may be of as broad a distribution.

ATOMIC PHYSICS

Atoms were initially thought to be the smallest unit of matter, but we now know there are subatomic particles, and we know of wave–particle duality in the conceptualization of physics. However, for the purposes of discussion of neurophysics, we start our discussion of atoms and their constituent protons, neutrons, and electrons, in the classic sense.

The features that we need to highlight in our discussion of atomic physics include just a few: charge, the ability to bond with other atoms, and the importance of isotopes and radioactive decay.

The structure of atoms that we think of is a construct which we create so we can envision them in our mind's eye. We know that planets are objects

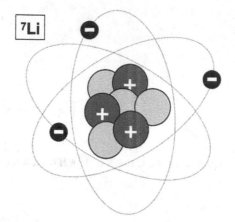

Figure 1.1 Atomic structure. Diagram of a lithium atom, with three protons and four neutrons. Lithium-7 is the most common and stable isotope of lithium.

which orbit stars, so when we think of an atom, we think of a nucleus at the center and concentric rings of electrons orbiting the nucleus, in much the same way. Of course, we now know that this is overly simple and not as good an analogy as we once thought, but for educational purposes, we can still consider a central nucleus and orbiting electrons. Figure 1.1 shows a simple diagram of a lithium atom. In general, lithium is not a pivotal atom for neurophysiological function, but this demonstrates atomic structure and has isotopes of differing stability.

HOW TO MAKE A UNIVERSE

The first element created after the big bang seems to have been hydrogen, and that was not even early in the celestial sense. When the energy of the early universe settled down to create primordial subatomic particles, they were independent, not bound, because the early time and space were so energetic. Eventually, the universe expanded, there was a reduction in local energy, and the early subatomic particles could interact and form the first hydrogen atoms—one proton and one electron. This gas phase was characterized by regions of inhomogeneity, which facilitated regional coalescence of early matter into clouds that then contracted into the first generation of stars. Gravitational attraction concentrated regions of randomly increased density, making the inhomogeneity even greater. As the first clouds became the first stars, the gravitational attraction pulled in even more hydrogen. Ultimately, there were huge balls of swirling gas with the highest density at the center. The pressures in the center of these first stars became immense. Collisions were frequent. Temperatures rose in the center due to these close atomic interactions. The high temperature and high pressures reached fusion threshold. Two hydrogen atoms would fuse to form a helium atom. With that transition was the production of even more energy. The star became hotter.

When stars use up their fuel, the outward pressure of the generated energy from fusion is lost, and the stars collapse. This collapse can result in such high internal pressures that the stars then explode, with some of the constituent elements fusing to make heavier atoms, each with more protons.

Fast forward to the next generations of stars, and we get to stars such as ours, in which there is more fusion, making heavier elements. A star about the size of our sun can fuse to form a nucleus of, at most, 26 protons—an iron nucleus. Our sun does not have the size and thereby the energy to produce heavier elements. But fortunately there are larger stars, and many of the important elements we discuss in this book are made when these collapse and explode—the self-destruction of the dying giant stars results in fusion of heavy elements, making even heavier elements.

That brings us to today, when our sun is still mostly hydrogen, and our planet is a combination of debris from previous stars.

OUR PALETTE OF ATOMS

In our discussion of the physics of neurology, there is a fairly limited set of atoms that we need to consider, and a fairly limited set of features that are important to our story. The atoms of life that we mainly think of are hydrogen, oxygen, carbon, nitrogen, calcium, sodium, chlorine, phosphorus, iron, potassium, sulfur, and magnesium. These together account for more than 99.9% of atoms of the human body. The first four account for more than 99% of the body's atoms.

The atoms are frequently represented in a periodic table of the elements, which we have the good taste of not showing in this text. The important point is the atomic structure of some of our key elements of interest.

For some of the discussions in this book, it is important to be aware of stability or instability of the atoms, the potential for decay, and for that we need to discuss the difference between atomic number and mass number—that is, the structure of isotopes.

The number of protons in an atomic nucleus is the *atomic number*, and this determines which element it is. The hydrogen nucleus has one proton. The number of neutrons in the atomic nucleus determines the *mass number*, which is equal to the sum of the protons and neutrons. If a nucleus has one proton and one neutron, it is still hydrogen with an atomic number of 1 but now the mass number is 2.

The elements are ordered by the atomic number, and when designating an element, the atomic number is generally a small number to the left of the symbol. For example, lithium has three protons, so the designation would be $_3$Li; however, because the name of the element is dependent on the atomic number, the presence of the symbol makes the indication of the atomic number irrelevant, so we usually just use Li or similar without the subscript. The atomic mass, counting protons plus neutrons, does change with the same element, so that is generally indicated as a superscript preceding the element symbol. The most stable and most prevalent lithium isotope is where there are four neutrons accompanying the three protons. So 3 + 4 = 7, which is the mass number, and the symbol would be ^{7}Li.

Bonding

The first property we are interested in is bonding of one element to another. We consider two types of bonds—covalent and ionic. *Covalent* bonds are where two atoms share electrons. These bonds can be difficult to break, taking significant energy to separate the atoms. *Ionic* bonds are where the atoms are in ionic form, having donated or accepted an electron with another atom. Each ion has a charge, and the differing charges attract the atoms, holding them together. Covalent bonds generally hold together tightly, although there is a great difference in bond strength. In general, the closer the atoms are to each other, the stronger the bond, or in other words, the higher the *bond energy*—this is the energy that would be needed to pull the atoms apart. For bonds where there are two or more shared electrons, such as in a double bond holding carbon to nitrogen or carbon to oxygen, the bond energy is much higher.

Charge

Substances held together with ionic bonds are bound by proximity and charge attraction, so they can more easily come apart, such as with table salt—sodium chloride, $NaCl$; this dissolves in water without having to add external energy to pull the atoms apart. Bonding is important for understanding ion dissolution in electrolyte solutions and for understanding bond dissolution from metabolism and some types of molecular damage.

Isotopes

Stability of the elements is determined by the number of neutrons and the total number of protons of the atom. Elements that are unstable exhibit radioactive decay. This decay can be alpha decay (where two protons are emitted, essentially a helium nucleus), beta decay (where an electron or positron is emitted), or electron capture (changing a proton to a neutron, thereby making the atom an element of atomic number one lower than the original atom). Although generalizations often break down and that is certainly true here, higher numbers of protons make an atom more unstable and more susceptible to radioactive decay. In contrast, generally speaking, more neutrons make an atom more stable.

So, we have some of the basic properties of atomic physics needed to talk about neurophysics:

- Atoms have positive and negative charge, and under certain circumstances they can donate a loosely held electron and so have a net positive charge or accept an electron into an orbital and have a net negative charge.
- Atoms bind to other atoms with differing energies. The difference in these energies and the conditions in which these bonds can be broken determine much of their chemical properties.
- Atoms come in isotopes with differing numbers of neutrons and differing stabilities. This predisposes them to decay, where they change to a different isotope or a different element.

Atomic theory is essential for understanding electronics, and atomic decay is essential for understanding much of the laboratory techniques that use radioisotopes as a method to detect substances with a specific characteristic.

CHARGE MOVEMENT

A critical aspect of physics that pertains specifically to neurology is charge movement. From the human body to neurodiagnostic equipment, there is physics behind the movement of charge. This section discusses current flow in a general sense. Specifics of equipment and diagnostic studies are discussed in subsequent sections.

Positively and negatively charged substances move in an electric field unless otherwise constrained. It is easier to illustrate flow of current in a conductor than in a liquid, and the physics is different. Thus, we first consider flow of charge in a conductor.

A conductor has the ability to conduct charge. In the case of metallic conductors, the charge movement is electrons moving through the material from the region with higher negativity to the region of lower negativity. This gradient moves the electrons from atom to atom.

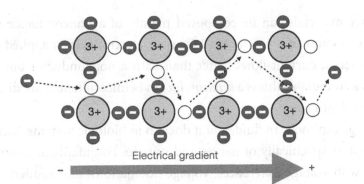

Figure 1.2 Diagram of a conductor. With three valence electrons, there is an empty spot in the electron orbital amenable to temporary location of an electron. As the electrical gradient drives electrons from left to right, the electron moves from atom to atom. Not shown is the fact that it is often not the same electron that does the moving: One electron approaches an atom, and that or another then leaves down the gradient.

Figure 1.2 shows a massively enlarged diagram of a conductor. In this circumstance, the valence of the material is such that although the material is electrically neutral, there are spaces in the electron orbitals (classically speaking) where electrons can reside, thereby filling an orbital. We say "classically speaking" because of the wave–particle duality of electrons and other particles, so the physics reality is more complex but is functionally as we discuss. Wave–particle duality and quantum physics are outside the scope of this book.

An electron enters from one side and then fills an empty spot. This is atomically stable although not electrically neutral. Subsequently, an electron leaves this atom and jumps to another open spot downstream in the electric field. It need not be the same electron that moves.

When we back up from this microscopic view and look at the whole conductor system, current is flowing through the conductor material down the electrical gradient.

Some materials conduct better than others. Substances that do not have any spots where electrons can visit are *nonconductors*. Unless there is so much applied electric field that there is a spark across the terminals, there will not be current flow through a nonconductor material.

Other materials can be composed mainly of a nonconductor with a smaller amount of a conductor. When an electric field is applied across this material, current flows better than with a nonconductor but not as well as a conductor; this is a *semiconductor*. Semiconductors are discussed more in-depth later.

Charge can move in fluid, and it does so in biologic systems, including our context specifically of nerves and muscle. Foundational neurophysiology with voltage differences, voltage fluctuations, and sudden change in conductance is responsible for the electrical properties of nerve and muscle. In fluid conduction, current can be carried by both positively and negatively charged atoms and molecules. Sodium and chloride in solution are the simplest examples of charge-carrying fluid constituents.

FLUIDS

Fluid dynamics is the physics of fluids as we consider them here, but the field also includes movement of gasses, which is not addressed in this book. The specific subfield of interest is *hydrodynamics*, which relates to the motion of liquids. For our purposes, we consider the classical hydrodynamic thesis of fluids being a relatively homogeneous substance, termed *continuous*. This means that for purposes of our discussion, the complexity of blood and other body fluids including the presence of cells and other inhomogeneous materials is not considered, but other factors of the fluids are considered crucial, especially density, viscosity, velocity, and pressure. A term that we should probably embrace is *neurohydrodynamics*, which is the specifics of fluid mechanics as it applies to neurologic systems.

Some of the areas of neurohydrodynamic concern include the following:

- Ischemic stroke
- Hemorrhagic stroke
  - Predisposition to hemorrhage
  - Vasospasm post-subarachnoid hemorrhage

- Venous sinus thrombosis
- Cerebrospinal fluid (CSF) circulation disorders
  - Idiopathic intracranial hypertension
  - CSF hypotension
  - CSF drains and shunts
- Polycythemia effects on neuronal circulation
- Use of hydrodynamic principles in select imaging studies
  - Cerebrovascular and cardiac ultrasound
  - Magnetic resonance angiography and venography

Ischemic Stroke

Vascular anatomy shows that arterial critical stenosis or occlusion produces loss of oxygen and nutrients to supplied brain tissue, with the expected manifestations of ischemic stroke. Because the arterial tree has connections that provide for alternative pathways, the deficit can be lessened or even abolished by collateral flow. The ability of collateral flow to adequately supply neuronal tissue depends on the richness of the vessels, degree of vascular pathology affecting those vessels, and timing of the resupply. For example, more than 30% of patients who have been diagnosed with transient ischemic attack have signs of acute ischemia on magnetic resonance imaging, indicating that there was complete clinical recovery but incomplete neuronal recovery.

Hemorrhagic Stroke

Among the characteristics of fluids is pressure, and each vessel in the vascular tree has a sensitivity to a certain level of blood pressure. The pressure is most pulsatile in larger proximal vessels, and there is less of a pulse effect with progressively smaller vessels and only a small amount in the veins. Increased pressure, both mean and peak levels, predisposes to cerebral hemorrhage.

Vasospasm

Vascular resistance is increased by cerebral vasospasm whether due to subarachnoid hemorrhage (SAH), reversible cerebral vasoconstriction syndrome (RCVS), or other cause. The most common cause of death with SAH is vasospasm rather than the bleed itself. The pathology of post-hemorrhage vasospasm is not completely understood, but it is thought to be due to effects of leaked blood directly on the exterior of the arteries. Added to the vasospasm of SAH is increased intracranial pressure, which together serve to reduce cerebral circulation and cause ischemia. RCVS is rapid onset of vasospasm that presents with acute headache which can suggest SAH. Associated with RCVS can be focal neurologic deficits, whether motor and/or sensory or language. Seizures can occur. The precise pathophysiology of the vasospasm is not known, but the presentation is dramatic.

Cerebrospinal Fluid Circulation Disorders

Multiple disorders affect CSF. In acute care neurology, the most common are idiopathic intracranial hypotension, low-pressure headache, cerebral venous sinus thrombosis producing reduced CSF reabsorption, and CSF infection or SAH also reducing CSF reabsorption as well as worsening CSF pressure due to an inflammatory response. The commonality of the increased CSF pressure scenarios is that there is reduction in arterial perfusion because the arterial pressure gradient is reduced due to elevated downstream pressure. Low-pressure headache usually develops after lumbar puncture, although there can be spontaneous leaks that develop in the cerebral or spinal dura. Low spinal fluid pressure results in sagging of the brain and meninges when upright in the sitting or standing position, which can produce severe head and spine pain. CSF shunts and drains are used when it is thought that patients may have increased CSF pressure as etiology for some of their deficits, especially in patients with intracranial hemorrhage, ventricular trapping by mass effect, cryptococcal meningitis, or normal pressure hydrocephalus. Note that shunts and drains can produce low pressure headache.

Polycythemia

Marked increased hematocrit can produce arterial and venous stroke. In some patients, the underlying cause of the polycythemia can also produce disordered platelet function predisposing to stroke, but in the context of this discussion, we are referring to the hemodynamic effects. Elevated blood viscosity produces increased resistance to flow, slower flow with more edge resistance (edge of the column of fluid against the vessel wall), and therefore reduced cerebral blood flow.

Diagnostic Studies

Several key neurodiagnostic studies rely on neurohydrodynamic principles for performance. Among these are ultrasound and the noncontrasted versions of magnetic resonance angiography and magnetic resonance venography. For ultrasound, high-frequency sound waves are used to identify and characterize vessel and cardiac flow. Structure is identified by the sound waves reflecting off of the interfaces of tissue densities. A border between a dense tough tissue (e.g., mass or arterial wall) will look very different from interstitial space or especially a fluid-filled vessel interior. These properties are essential for these diagnostic studies.

Basic Electronics

KARL E. MISULIS AND EVAN M. JOHNSON ■

HISTORY OF ELECTRONICS

Electronics have been used for centuries. We do not know when someone first noticed static electricity, but we have better documentation of when the first static electric generators were produced. Otto von Guericke was a true scientist who first became known in the 17th century for his study of vacuum physics, but then became even better known for the study of static electricity. The static phenomenon was known about since antiquity, but this was not known to be due to separation of charge. Von Guericke made static generators that largely relied on friction principally of nonconductors to separate charge.

The next major advance occurred when Alessandro Volta invented the battery in 1799. This provided a reliable and somewhat steady supply of current with which to create and use electronic circuits. Among his observations was the fact that electrical potential and charge were proportional. The unit of electrical potential became the *volt*.

Georg Ohm used batteries of Volta's design to further study electricity and found that there was a direct proportional relationship between the amount of potential difference and the resultant flow of electric current.

This culminated in Ohm's law, where the voltage is equal to the current multiplied by the resistance:

$$V = I \times R$$

Over the ensuing many years, the principles of electrical current, voltage, conduction, and resistance were elucidated and characterized. Circuits became more complicated. The age of electronics had begun. These were analog electronics, with the term *analog* deriving from the Greek word for "proportional." This is an appropriate designation because the analog signal does not have the incremental steps of digital technology.

Advances in electronics developed through scientific investigation and ultimately practical utility. Edison was only one contender in the race to produce efficient and sustainable electric lighting. Nobel Prize winners Bardeen, Brattain, and Shockley of Bell Labs were not the first to propose what we now call the transistor; what we now call a field-effect transistor had been proposed a generation earlier.

Subsequently, transistors enabled arrays of circuitry, which enabled efficient logic gates, fundamental for digital electronics. Logic gates are circuits that provide output not of a continuously variable magnitude but, rather, of only two possibilities—on or off, or 1 or 0. Tube electronics could be used for gates and was key to some of the first computers, but tube failure and tremendous physical and energy requirements made them impractical. Integrated circuits with transistors, resistors, and capacitors were essential for entrance into the digital age.

Nowadays, almost all of our electronics are digital. There are some analog amplifiers in our neurophysiologic equipment, but these are generally designed to raise the very small biologic signals to a magnitude that can be digitized and then handled digitally for display and analysis.

Next, we consider electronics in more detail, beginning with the most basic circuit element—the conductor.

CONDUCTORS, NONCONDUCTORS, AND SEMICONDUCTORS

Material can be classified as conductors, nonconductors, and semi-conductors, depending on their ability to conduct charge.

Conductors are so called because they are composed of materials that have the ability to conduct electricity. We hope that this discussion does not give too much indigestion to quantum physicists, but we can generally say that conductors have an atomic structure that allows electrons to move through the matrix. We can envision conductors such as copper having loosely held electrons that can move through the material. If we apply a charge differential across a piece of copper, the electrons on the negative side of the copper can flow through the matrix. Electrons flow down the potential gradient from negative to positive. This is a conduction.

Nonconductors are so called because they do not allow electrons to flow easily. The electrons are tightly bound to the associated atoms and cannot move from one atom to another. If a potential difference is applied across a nonconductor, then electrons cannot move through the material, unless so much potential difference is applied that there is a spark.

Semiconductors are so called because they semiconduct. They conduct better than nonconductors but less well than conductors. These are generally made of materials such as silicon that are nonconductors with a bit of conductor mixed in, called *doping*. This material might be arsenic or gallium or other material. Figure 2.1 shows a nonconducting silicon matrix and two semiconductors created by doping the silicon with other elements as detailed below.

The magic of semiconductors is gating of flow when different types of semiconductors are in conjunction. Semiconductors differ depending on elements with which the nonconductor material is doped. Let us consider semiconductor materials with a silicon foundation. Silicon has four valence electrons. Valence refers to the ability of an atom to interact with other atoms by sharing an electron, donating an electron, or receiving an electron. Silicon has a valence of 4, so it can interact with four nearby

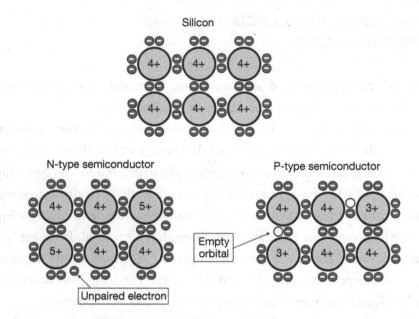

Figure 2.1 Semiconductors. Silicon matrix is a nonconductor. When doping with a pentavalent element, there is a weak held electron that is available to move. On the other hand, if the matrix is doped with a trivalent element, there is a spare hole in which an electron can temporarily reside.

atoms. Typically two silicon atoms mutually share an electron from each, thereby the two electrons fill an electron orbital.

The concept of stability of filled electron orbitals being a stable atomic configuration certainly is easy to conceptualize and does explain much of the behavior of atoms. However, we now know that this is not completely correct. Electrons are not simple little particles. They have features of particles and waves—the wave—particle duality mentioned in Chapter 1. When we try to understand the function of something we cannot see, we try to make an analogy we can visualize in order to better understand the concept. But all macroscopic analogies fall short. Suffice it to say, when it is convenient to consider electrons to be particles, we do so; otherwise, we consider them waves.

We now return to our discussion of valence and semiconductors. Silicon forms a matrix with surrounding silicon atoms, each sharing a valence electron with an adjacent atom. This means that there are no more electrons available to move through the matrix. Silicon is a nonconductor. Because arsenic has a valence of 5, when we insert a small amount of arsenic into the silicon matric, four of the electrons pair with electrons from nearby silicon atoms, leaving one remaining electron that is not paired. This is a loosely held electron. With the electron hanging around the arsenic atom, the region is electrically neutral, but with a little push, that electron can wander through the matrix. This push is an electric field, in our visualization.

Alternatively to arsenic, we can put a small amount of gallium into the silicon matrix. Gallium has three valence electrons, which means that it can interact with three nearby atoms. This leaves one hole or spot where an electron can sit. Although not electrically neutral, it is atomically stable. The electron is loosely held and can move from atom to atom.

The semiconductor doped with a material that has left it with an open slot for an electron is termed *P-type* because it acts like a positively charged material; it wants an electron for the open spot. A semiconductor doped with an element that has an additional electron is termed *N-type* because the added electron is loosely held and therefore is like a negatively charged material, willing to give up an electron.

Semiconductors leverage the difference between the valences of these atoms to create variable conductance where the variation is due to gating by an applied potential difference. Semiconductor applications are discussed in more detail below.

CIRCUIT ELEMENTS

Resistors are elements that resist the flow of current. They conduct but dissipate some of the energy of the electrons usually as heat, although sometimes as light. The amount to which the resistor dissipates energy is the *resistance*. Figure 2.2 shows a resistor in series with a power supply. The energy of the power supply is dissipated across and resistor, and most of the energy is converted into heat.

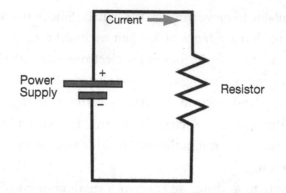

Figure 2.2 Resistor. A resistor is in series with a power supply. Current flows clockwise; positive current by convention. Electrons are moving counterclockwise. Energy from the power supply is dissipated by the resistor usually as heat.

Capacitors are elements that store energy by separation of charge. As opposed to batteries, they do not generate the charge differential chemically. They are composed of two sheets of conducting material separated by a thin layer of nonconducting material. The terminals are wires connected to each of the sheets. Figure 2.3 shows a diagram of a capacitor connected to a power supply. When a power source is connected to the terminals of a capacitor, electrons flow onto of the layer connected to the negative end of the power supply and electrons flow off of the layer connected to

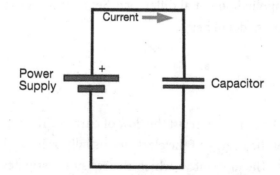

Figure 2.3 Capacitor. A capacitor is in series with a power supply. When the power is on, current flows clockwise, and electrons flow counterclockwise. During this flow, the capacitor is charged. The charge built up across the capacitor opposes the further flow of charge, which stops when the charge is equal and opposite to that of the power supply.

the positive end. As this happens, a potential difference is built up across the capacitor. If there is no change in the applied potential difference, the potential difference across the plate of the capacitor gets high enough to ultimately oppose any further movement of electrons—the capacitor is charged, and no more electrons move. The amount of charge that the capacitor can handle is the *capacitance*. If the applied electrical potential is turned off but the capacitor terminals are still connected, then the electrons gathered on the negative side of the capacitor flow off of that layer, through the circuitry, and onto the opposite layer, discharging the capacitor. This flow of current is *capacitive current*.

Inductors are circuit elements composed of a coil of wire with two terminals. If one end is connected to the negative terminal of a power supply and the other connected to the positive end, current flows through the wire. A circuit diagram of an inductor and a power supply is shown in Figure 2.4. When current flows through a conductor, there is a small magnetic field oriented radially around the conductor. Usually, the magnetic field is small and essentially trivial. However, when the wire is coiled, the magnetic field produced by the current is nontrivial, and a significant proportion of the energy of the potential difference is used up in generation of the magnetic field. This field is oriented according to the geometry of the coil. Eventually, the magnitude of the magnetic field becomes stable for

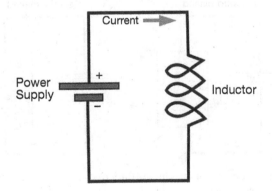

Figure 2.4 Inductor. An inductor is in series with a power supply. When the power is engaged, there is current through the inductor. Energy is stored in the magnetic field of the inductor.

the amount of applied current. When the potential difference is switched off, the magnetic field collapses, and by induction, current is produced in the leads of the inductor in the same direction that current was previously flowing. When the magnetic field has totally collapsed, current ceases.

Diodes are composed of two semiconductor materials placed in a layered configuration. This results in electron flow only in one direction. These are used for rectifiers and gates. The details of the diode are important to understanding how transistors work, so we briefly discuss diodes.

Figure 2.5 is a diagram of a diode. Diodes are composed of two small bricks of semiconductor, one N-type and one P-type. When these are placed together, electrons flow as described, from the N-type into the P-type, making for a junction potential in the absence of a power supply. In some ways, this is similar to the membrane potential in neurons that develops from a potassium gradient. Confusingly, the junction potential is oriented such that the N-type material has a positive charge and the P-type has a negative charge because electrons moved from N-type to P-type materials.

When a power supply is attached to the diode, the result can be good flow or no flow. If the orientation of the power supply is such that the positive terminal is attached to the P-type material and the negative terminal

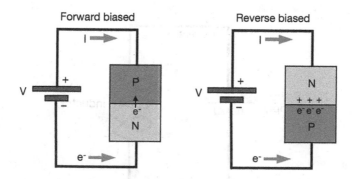

Figure 2.5 Diode. Diode can conduct in one direction. Depending on the orientation of the semiconductors, the interface would be forward biased and current flows; otherwise, it is reverse biased and the potential difference only served to augment the junction potential with no net current flow.

is attached to the N-type material, current can flow. This is because in achieving steady-state diffusion potential, sufficient electrons wandered from N-type to P-type until a charge differential across the junction opposed further electron migration. The power supply is in the opposite polarity to the junction potential, and current flows readily through the diode, although not as freely as with a conductor.

If the N-type and P-type materials are flipped so that the positive terminal is attached to the N-type, then the power supply augments the junction potential, further impeding current flow. So the first example (Figure 2.5, left) is forward biased, allowing current flow, and the second example (Figure 2.5, right) is reversed biased, opposing current flow.

Transistors are composed to two or three pieces of semiconductor with typically three terminals. Flow from one terminal to another terminal is gated by charge applied to a third terminal. Details of the function of transistors are beyond the scope of this book, but suffice it to say that the gating mechanism can be to inhibit or enhance current flow. Transistors are discussed further in the Amplifiers section.

We are seeing a few themes here. Electrons flow through various substances, producing current. All substances, even superconductors, have at least a little resistance, so some of the energy of the moving electrons is dissipated as heat or light. Energy flowing through conductors also produces a magnetic field around the conductor that is almost trivial unless the wire is coiled, lining up all of the magnetic fields, or unless the current flow is large. So, magnetic fields as well as electrical potential differences can produce current flow. Last, the electrical and even electromagnetic effects of current flow through these elements can be made to accomplish amplification, filtering, rectification, and other electrical functions.

ELECTRONIC COMPONENTS

This section discusses analog circuitry. Chapter 3 discussed computers and applications as they pertain to neurology. We begin with discussion of amplifiers and then briefly discuss analog filters. Subsequently, analog-to-digital converters (ADCs) are discussed.

Amplifiers

Our physiologic signals need amplification in order to be analyzed and visualized. The microvolt and millivolt potentials that are typically recorded from electrode leads are too small to drive cathode ray tubes and even too small to serve as inputs to the ADCs required to get the signals into the computers for data manipulation, analysis, and display.

Amplifiers use an exogenous power source. Because ±1 mV is not enough signal for an ADC, we would need to amplify the signal at least 1,000 times so that the signal is ±1 V. Historically, a tube amplifier would have been used.

Tube amplifiers work by using sealed vacuum tubes that have three connections—cathode, anode, and grid. A power supply is connected to the cathode and anode, which causes the cathode to heat up, exciting electrons to jump off of the cathode and onto the anode. Current flows and the amount of flowing current is measured by a device between the tube and the power supply. A signal is applied to the grid, and very small changes in grid voltage can greatly change the ability of electrons to jump from the cathode to the anode. This gating results in large swings in current flow, which are then translated into changes in voltage downstream from the tube. The changes in output voltage are far larger than the changes in grid voltage.

Transistor amplifiers work through a similar gating mechanism, although the physics is different. A power supply is placed across two leads of a transistor and the signal is applied to the third. Figure 2.6 shows a negative–positive–negative (NPN) bipolar junction transistor. For this NPN transistor, there are two N-type semiconductor wafers separated by a thin P-type layer. The terminal on one of the N-type segments is the *collector*, the terminal on the other N-type segment is the *emitter*, and the terminal on the P-type is the *base*.

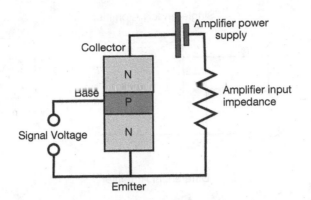

Figure 2.6 Transistor: Transistor as part of an amplification system. A small signal voltage across the base emitter junction gates the flow through the right side of the circuit, where a larger power supply drives a large current that produces marked current flow. The voltage drop across the amplifier input impedance is much larger than the signal voltage.

In Figure 2.6, the base–emitter junction is forward biased in that current can flow relatively easily in that direction. On the other hand, the collector–base junction is reverse biased in that the applied voltage serves to increase the junction potential at that junction, thereby blocking electron flow. So with no applied base–emitter potential, there is little current flow. Whereas the base–emitter junction can conduct, the collector–base junction cannot.

When a small voltage is applied at the base–emitter junction, represented as V_{BE}, this can be considered to remove some of the electrons at the collector–base junction, allowing for there to be holes in the base material for current to flow from emitter, through base, to collector. In this fashion, a small amount of V_{BE} voltage controls current through the V_{CE} transistor circuit. We can see how this effect can be used to amplify a signal. The small biologic current applied through the base–emitter junction results in a much larger collector–emitter current, hence amplification.

This amplifier system performs and amplifies well, but we need one more component—that is, to reduce the noise in the system. It is typical to have common noise in our electrode leads. Because of this, we use differential

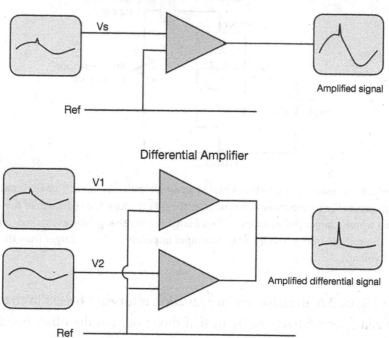

Figure 2.7 Amplifier systems. (*Top*) In a single-ended amplifier, the output is identical to the input except for amplitude. (*Bottom*) In a differential amplifier, the low-frequency noise is present in both leads and is partly suppressed by the differential amplifier while the signal of interest is amplified.

amplifiers. Figure 2.7 shows single-ended and differential amplifiers. The single-ended amplifier produces a faithful representation of the original signal, artifact as well as biologic signal. The differential amplifier produces an amplified version of the unique signal, eliminating the contamination by electrical artifact—in this case, a slow rolling sine wave. This is accomplished by subtracting the amplified signal of a reference electrode from the amplified signal of the electrode with the potentials of interest.

For most of our circuits, the gain from this amplification is 10× or less. So, to produce much larger amplification, the transistors are ganged. If each stage produces a 10× amplification, then three stages produces a 1,000× amplification (10^3).

Filters

Filtering is needed because the data that we review for each neuro-physiologic study present in a certain time course and in a certain frequency band. The concept of frequency bands applies not just to rhythmic activity but also to single waves, There is a collection of waves which together comprise the signal of interest. Activities higher or lower than this window are likely noise or otherwise unrelated. We will see the signals of interest better if we remove those frequencies. This can be done by analog or digital methods, which are discussed later.

Here, we emphasize that the filter actions can be complex. Filtering alters the shape of a complex signal, and without appropriate care, the signal can be so deformed as to distort the wave to either make it uninterpretable or give the wrong answer.

We should consider some of the effects of changing filters on a particular signal. This is hypothetical, but it certainly applies to our setting of filters outside of established settings for our instruments.

Say that we do a recording and it has a complex potential that has some very fast frequency artifact. If we set the high-frequency filter to a lower frequency, then the amount of high frequency is reduced. However, the amplitude of our signal of interest is also reduced because the high-frequency filter is not an absolute frequency break and also there is a faster component to even our slower potential of interest. Similarly, the duration is increased because of the loss of that fast phase.

On the other hand, let us consider a study in which we are mainly interested in the fast components but yet there is a slow-rolling wave superimposed upon our recorded signal. If we turn the low-frequency filter to a higher level, we certainly will reduce that slow wave contamination. However, we will also reduce our signal because although it is mainly fast, it does have a slower component. Also, we reduce the latency because we have removed some of the slower component of our signal of interest. The signal appears sharper than it would have otherwise.

The message from this is that it is best to try to eliminate fast and slow artifact by optimal technique rather than overuse of filters which can distort waveforms. We always have to use filters, but we try to minimize the effect that adjusting these can have on our recording.

Physiologic Device Components

Physiologic recording of neuroelectric activity necessitates electrodes, which are the bridge between the body and the electronics. Electromyography (EMG) and evoked potentials (EPs) use stimulating and recording electrodes, whereas electroencephalography (EEG) uses only recording electrodes.

Skin electrodes are composed of a variety of metals. Silver and gold have been used prominently in the past; other metals are now more commonly used. Electrodes are described as reversible or nonreversible. Reversible electrodes have bidirectional chemical reactions so that charge moves freely. Nonreversible electrodes have impaired current flow in one direction, whereas current flow in the opposite direction is facilitated. For most neurophysiologic recordings, this is not a major issue because the amount of current flow is small.

The leads from the electrodes to the physiologic equipment are not simple conduits. They have some resistance and capacitance. Resistance develops when the leads are long or when there are breaks in some of the elements of the multistranded wire. Capacitance develops when leads are in close proximity. This is important because when there is resistance and capacitance in a loop circuit, we have a resistor–capacitor (RC) circuit (Figure 2.8).

In Figure 2.8, the subject of the recording is the source of power on the left and the recording equipment is on the right. The resistor symbol is not an actual resistor but, rather, the resistance of the electrode wire and the electrode–patient interface. The capacitor is not a circuit element but, rather, represents the capacitance of the leads in close proximity. Because of the filtering effect of RC circuits, the electronic signal

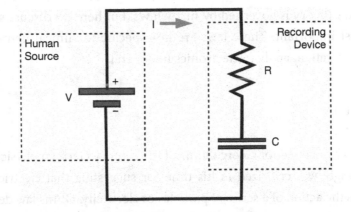

Figure 2.8 Virtual resistor–capacitor (RC) circuit in physiologic recording.
Electronically, the RC circuit is connected to a power supply. Current flow charges the
capacitor. The time to charge the capacitor and stop flow is proportional to the resistance
and inversely proportional to the capacitance. In this discussion, we are talking about
a virtual circuit with real implications. *V* is potential from the patient, and R and C are
resistance and capacitance of the recording system, respectively. The recorded signal is
distorted by the circuit.

seen on the right side of the diagram will be distorted. The input end
of our recording device is looking not only at the potential difference
between the electrode leads but also the potential difference across the
virtual capacitance. Faster frequencies can be shunted by this capacitance, moving back and forth so quickly that the device does not see
those high-frequency changes. So the unwanted RC circuit works like
an unwanted high-frequency filter.

We minimize this distortion by having the resistance component low
and the capacitance low.

CIRCUIT THEORY

We previously discussed some of the elements of circuits, but the concept
of a circuit itself is a bit complex. A circuit is a series of connections and
elements that result in signal flow. For most electronic circuits, this is an
analog signal—a continuously varying voltage through a conductor and
other circuit elements.[1]

Circuit theory is governed by many laws, but here we discuss some of the most important. These laws are not specific to clinical neurophysiology but, rather, apply to electronics in general.

Ohm's Law

Ohm's law is named for Georg Ohm, a 19th-century German physicist who interestingly was criticized in his time for suggesting that electrical current was the action of a series of particles of electricity. Ohm's law describes the relationship between voltage, current, and resistance for a specific type of circuit. Figure 2.9 shows a resistor in series with a power source. These elements could be physical electronic components or could be parts of biologic tissues, meaning that the resistor could be resistance of body tissues and power source could be potential difference across cellular membranes.

Ohm's law states that for a resistive circuit, there is a direct relationship between imparted power and current flow. This is intuitive, as we increase the voltage, more current will flow:

$$V = I \times R$$

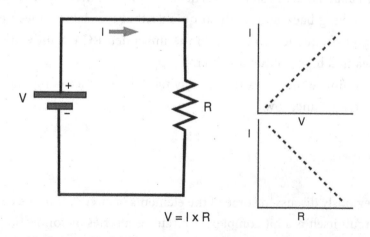

Figure 2.9 Ohm's Law. Ohm's law describes the relationship between voltage, resistance, and current. The small graphs on the right show the relationship. Additional details are provided in the text.

If we rearrange the variables, we can get

$$I = \frac{V}{R}$$

This formula states that current is equal to the voltage divided by the resistance. This might not seem intuitive, but it make sense that as voltage increases, the current will increase, and if resistance increases, the current will decrease.

Although this formula does not need an analogy, one that we often use and is illustrative is to imagine a pail of water. The pail has a hole in the bottom, so water leaks out at a slow rate. The height of the water in the pail corresponds to the voltage. The higher the height of the column of water, the greater the pressure and therefore the greater the water flow.[2] So the flow of water (current) is directly proportional to the height of the water (voltage) and indirectly related to the resistance to flow (smallness of the hole). Later, we return to this water and pail analogy to illustrate a more complex relationship.

Kirchhoff's Laws

Kirchhoff's laws are from Gustav Kirchhoff, a German physicist also of the 19th century, although later than Ohm. He is best known for his description of black body radiation,[3] but here we are interested in some of his circuit laws. Figure 2.10 shows the diagrams needed for this discussion—the top portion for the current law and the bottom portion for the voltage law.

Kirchhoff's current law states that the sum of currents flowing in and out of a node or connector point is zero. A node is where at least two wires connect and has no ability to store or modify charge. Therefore, all the current flowing in one wire has to be leaving by the others or some combination thereof. This seems intuitively obvious, but it is an important point. The formula is

$$\pounds_i = 0$$

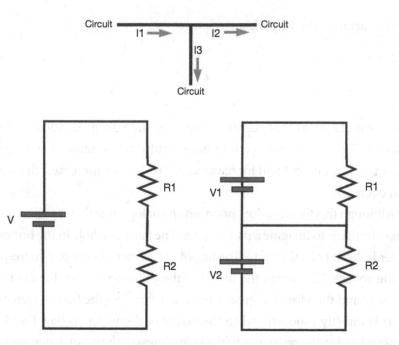

Figure 2.10 Kirchhoff's laws. (*Top*) Kirchhoff's current law, stating that all current entering a conductor node leaves. (*Bottom*) Voltage sources and drops balance across all resistive circuits, including the simple circuit on the left and all three loops on the right—the upper and lower loops and the large loop encompassing both power supplies and both resistors.

Kirchhoff's voltage law is less intuitive than the current law. The voltage law states that for any circuit loop, the sum of the voltage sources is equal to the sum of the voltage drops. Voltage source is any power supply or other element that can add electrical energy to the circuit. A voltage drop is any element that dissipates electrical energy. A battery is a voltage source. A resistor is a voltage drop.

Figure 2.10 (bottom right) shows a slightly complex circuit in which there is a power source with two connected resistors. Kirchhoff's voltage law applies for each of the small loops and also for the big loop encompassing both resistors and both power supplies.

Kirchhoff's laws are used to describe some basic circuit properties.

Application of Circuit Laws

Circuit laws can be used to describe physical properties of electronic circuits. Two simple applications are the calculation of equivalent resistance when resistors are in series or in parallel.

SUMMATION OF RESISTORS IN SERIES

The left side of Figure 2.11 shows a diagram of two resistors in series, connected to a power source. The resistances of the elements are R_1 and R_2. What is the equivalent resistance of these resistors in series? Current has to flow through both resistors, so the total resistance is equal to the sum of the individual resistances. That is,

$$R_T = \Sigma R_i$$

where R_i is the individual resistances. This is intuitively obvious.

SUMMATION OF RESISTORS IN PARALLEL

Calculation of the equivalent resistance of resistors in parallel is not so obvious. The right side of Figure 2.11 shows a diagram of two resistors in parallel with a power source.

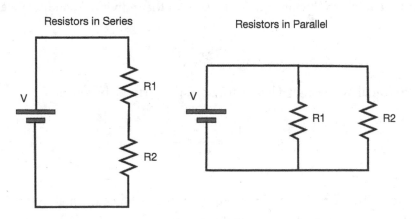

Figure 2.11 **Resistors** in series (*left*) and parallel (*right*). See text for details.

In this case, there are two pathways from one terminal of the power source to the other: One is through the pathway with R_1 and the other is through the pathway with R_2. Is the total resistance higher or lower than the individual resistances? Total resistance is lower because there are two pathways for current to flow. We can return to our analogy of a pail of water. If we have a pail with an intact bottom, resistance is infinite and water cannot leak from the pail. If there is one small hole in the bottom of the pail, there is a small amount of water flow through this high-resistance pathway. If we put another hole in the pail, there are now two pathways and more water flows. These are resistors in parallel. More resistors in parallel lower total resistance.

We calculate the equivalent resistance by considering the opposite of resistance, which is *conductance*. Conductance is the ability of a pathway to permit flow, which is the reciprocal of resistance. The letter symbol for conductance is G, so for the resistors,

$$G_1 = \frac{1}{R_1}$$

$$G_2 = \frac{1}{R_2}$$

And the total conductance G_T is the sum of the individual conductances:

$$G_T = G_1 + G_2$$

So to calculate the equivalent resistance, the formula becomes

$$\frac{1}{R_{eq}} = \frac{1}{R_1} + \frac{1}{R_2}$$

Or, verbally, the reciprocal of the equivalent resistance is equal to the sum of the reciprocals of the individual resistances.

ELECTRONIC DEVICES

We are gradually moving toward leveraging the physics and electronics to performing neurological diagnostic tests, but we have one more layer to build. This is the design and performance of electronic devices. Here, we consider how electronics work, and later chapters discuss how they are used for specific diagnostic and therapeutic purposes. Computer architecture is discussed in Chapter 3.

Many neurophysiologic studies depend on recording a response from the body, usually neuronal tissue, but sometimes muscle. The neuronal tissue can be in the central or peripheral nervous system, and for some studies, such as Evoked Potentials (EP), both central and peripheral recordings are made and used for interpretation.

Many neurophysiologic studies require a stimulus, such as nerve conduction study (NCS), and EPs. Others are purely recording, such as many epochs of EEG recording, although photic stimulation during EEG is routine. Electromyography (EMG) relies only on physiological activation, not exogenous stimulus. The relationships between device elements can be described by a series of block diagrams, in which the blocks are connected by data flow vectors.

A purely recording diagram, as for EEG, is shown in Figure 2.12. ADC is the device that turns analog biologic signal into digital data which can be calculated by the computer.

A stimulus and recording diagram, as for EMG and NCS, is shown in Figure 2.13. There is a stimulator arm of the circuitry, and this is driven by the same computer that does the acquisition analysis.

We now discuss some of these elements individually in more depth.

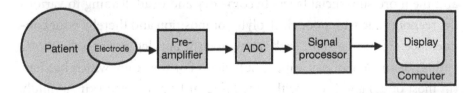

Figure 2.12 Data relay diagram for an **EEG machine**. ADC, analog-to-digital converter.

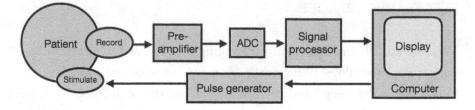

Figure 2.13 Data relay diagram for an **EMG machine**, which differs from EEG mainly in that there is a stimulator arm. ADC, analog-to-digital converter.

Electrodes and the Electrode–Patient Interface

Most neurophysiologic equipment uses electrodes of some sort. EEG and some EMG electrodes are disk electrodes. EMG also uses needle electrodes. The physical and chemical properties of the electrodes are often overlooked when considering the clinical systems.

Disc electrodes are an extension of the leads, designed to detect and conduct the biologic signals optimally. In general, this means a relatively low-resistance connection. To bridge the disc electrodes to the skin, a conductive gel is used. The gel facilitates ionic conduction and allows for some inevitable movement of the electrodes on the skin. In the case of electrocardiography, the conductive material, which is also adhesive, is part of a pad, with the metal portion of the lead embedded. For EMG/NCS surface leads, a small amount of electrode gel is used to bridge the otherwise high-resistance interface between skin and wire. For EEG, a paste is used after the scalp has been prepped by mild abrasion to reduce impedance by removing some of the dead skin and oils.

Skin preparation is needed for many neurophysiologic applications. This is because the skin is constantly growing new layers from below and the more superficial layers become dry and dead, leading to various degrees of tissue with poor electrolyte composition and thereby poor conductance. For some skin, such as many arm and leg surfaces for EMG and NCS, an alcohol pad wipe is sufficient. For the scalp, which has hair (in most of us) and is oily with much thicker layer of dead skin, a mildly abrasive material is used first to prepare the skin, and then a conductive

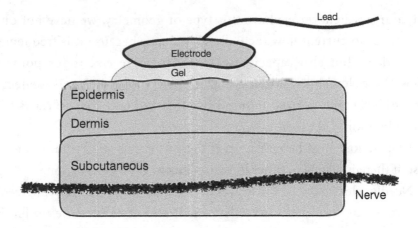

Figure 2.14 Skin and the electrode–skin interface. Electrode is connected to the irregular skin by a thin layer of conductive electrode gel. Not evident in the figure is that the upper levels of dermis and especially the epidermis undergo physiologic changes that make them less conducive to current flow.

gel is used beneath the cup electrode (Figure 2.14). The progressive change in skin physical and electrochemical function from base to skin surface is commonly termed *cornification*, and the physiologic name is *keratinocyte differentiation*.

Skin changes are of two basic forms. Deep layers have an energy-dependent apoptosis that results in preservation of the membranes while much of the machinery of the cell is ejected. Then, at superficial layers, this evolves to the cornification with which we are more familiar, where there is programmed cell death and formation of keratin filaments that form the structure of the dying cell. Part of this is an envelope forming around the cell. Last, there is intentional weakening of the cell–cell adhesion, which then allows for the outermost cells to be lost without damaging surrounding and underlying tissues.

The point of this complex discussion is that the skin transitions from a very moist conductive tissue with structure that allows current to flow easily to a structure that does not conduct electricity well at all. Removing some of the most superficial cells is essential for at least EEG. And for NCS, there is a need to bridge the gap between the metal electrode and the skin in the most conductive method possible.

Of course, whenever we have this type of geometry, we have not only resistance to current flow (actually impedance because it is frequency-dependent) but also capacitance. And from the electrode's point of view, these elements result in change in signal appearance. This effect is unavoidable, but we need to minimize this and take these physics into consideration.

Preparation must be enhanced if there is excessive cornification and also if there is edema or other impediments to good conduction.

Needle electrodes are used almost exclusively for EMG. Needle electrodes are insulated with only a small area at or near the tip for recording. The remainder of the electrode can be used for reference, or there can be a separate disc electrode on the skin for reference. Historically, needle electrodes were sometimes used for scalp EEG to minimize some of the signal distortion due to instability of the disc electrode. However, this has largely been discarded because of the risk of infection, disease transmission, pain to the patient, and the lack of necessity.

Electrode–Amplifier Interface

The communication between the electrode and the input side of the equipment may seem simple, but it is slightly complex, and thus there is opportunity for error generation.

Figure 2.15 shows a diagram of the electrode–amplifier interface. At the top of the figure is an idealized connection in which there is an electrode with a certain resistance (or more accurately impedance) recording a signal voltage from the body (V_s). In this diagram, we have separated the resistance and the voltage source, but in actuality they are combined. Nevertheless, the circuit diagram can be useful.

The right side of Figure 2.15 shows the box of the amplifier, which has two terminals for connection of the active and reference electrode leads. There is connection between these through the amplifier, although with very high resistance, so that there is limited current flow through the circuit. In order for the amplifier electronics to not significantly affect the

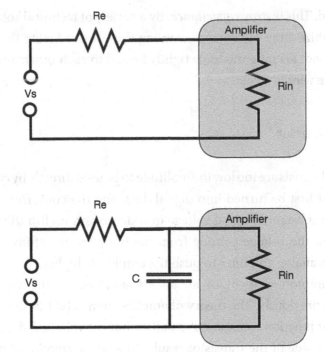

Figure 2.15 Electrode–amplifier interface. (*Top*) The ideal circumstance in which the amplifier input impedance sees the potential with some voltage drop over the resistance of the electrode leads and physiologic junction. (*Bottom*) A small stray capacitance is present, which then creates an RC circuit and the potential for altering the signal frequency components. Re is electrode impedance. Rin is amplifier input impedance. Note that while we speak of *resistance* for ease of calculation, we are really talking about *impedance*.

waveforms, it has to be high resistance. Again, we use the term resistance but we really mean impedance.

The bottom of Figure 2.15 shows one additional circuit element, a small capacitor. This is not a physical capacitor but, rather, a low level of capacitance between the leads coming from the electrodes. The implication of this capacitance is that in conjunction with the resistance of the electrode–body interface, this stray lead capacitance creates an RC circuit.

The result of the RC circuit is the making of a filter. Because the input impedance of the amplifier is bridged by the capacitance of the RC circuit, it is seeing a voltage from which some of the faster frequencies have been attenuated. This is *stray capacitance.* By a variety of technical solutions, we can minimize stray capacitance. Among these are reducing the length of the leads, not keeping the leads tightly bound to each other, and also not coiling the wire.

Analog Amplifier

Biological signals are too low in amplitude to be used directly by computers. They must first be turned into digital data, and the converters for doing so require substantial signal voltage in order to accomplish this task—far larger than the voltage coming from the body for any study. Therefore, there is an analog amplifier to push the amplitude higher.

The common element of an analog amplifier is the transistor, as described previously. The theory of amplification is best understood with the bipolar junction transistor. A small amount of voltage difference at the controlling side of the transistor results in a larger amount of voltage on the controlled side, and that amplification takes energy, by means of the power supply.

The concept of ideal amplifier is commonly discussed when considering electronic theory, where the output is an exact representation of the input but with a larger amplitude. In reality, no amplifier is ideal, although amplifiers may approximate an ideal amplifier over a specific range of voltages and frequencies. For a specific transistor amplifier, there are small low-amplitude fluctuations in input that will not produce a significant change in output. In addition, there are fluctuations in input voltage that are outside of the working limits of the amplifier, so there is little increase in output amplitude with a change in input amplitude. An amplifier has an input range, an output range, and a quantification of linearity over a certain level of input. An amplifier is used with an input and output range that has the best linearity.

There are other features of amplifiers that we might consider, including limitations on the frequency range of the input and output signals (bandwidth), efficiency of the amplifier (the relationship between power output and power consumption to produce that power output), and a series of properties that determine how well an amplifier deals with a change in voltage. This and noise we do not consider in more detail here, but it is important to know that amplifier function is complex, with limiting features and performance specifications.

Amplifiers are integral to analog filters. Although some analog filters use conductors, resistors, and capacitors (passive filters), most use amplifiers in circuits that have frequency-dependent transmission through the amplifier. In reality, all transistor amplifiers have frequency dependency, but when used for amplification purposes, the parameters of the input and output are such that nonlinearity and frequency dependence within the bounds of the analyzed signal bandwidth are minimal.

Analog-to-Digital Converter

We have reviewed how to take a very low-amplitude biologic signal and amplify it so that we can manipulate it and then use the data. For virtually all neurophysiologic equipment, the signal is transformed into digital data so that mathematical manipulation of the data can be performed. The first step, therefore, has to be analog-to-digital conversion with an ADC, as shown in Figure 2.16.

The signal must have sufficient amplitude so that the ADC can handle it. Microvolt inputs will not work for most ADCs; they usually work with signals of 1 V or more. So after the analog amplification and signal preprocessing are performed, the signal is fed into the ADC.

ADCs are used for almost any signal. It could be electrical potential, as for EEG or EMG; sound, as for digital music recording or voice recognition; or digital video capture.

There are several different types of ADC architectures, and details of all of these are beyond the scope of this book. However, discussion of two of these might be of interest. One compares the signal voltage to a ramp voltage that is generated by the device. The ramp has a known

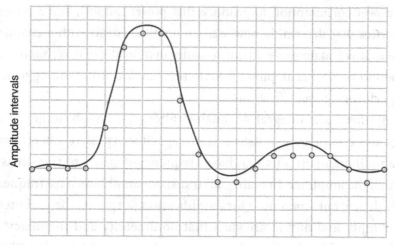

Figure 2.16 Analog-to-digital converter. Analog-to-digital converter action is essentially shown by placing a piece of graph paper over a biologic signal, here simulated for effect. Note that if we connect the dots, the resulting trace would be similar, but not identical, to the source. Of course, time and voltage resolutions are much denser in reality, making the display a faithful representation.

voltage and rate of change. The signal input voltage is compared to the ramp voltage, and when the signal voltage is equal to the ramp voltage, the time associated with the specific voltage is stored by the ADC, which generated the ramp. The binary equivalent of that voltage is stored, and then a sample is usually taken of that signal again or sometimes the ADC moves on to sampling the signal coming from the next biological channel. It takes far less than 1 msec for an ADC to determine a sample voltage.

An alternative method is for the ADC to make successive approximations of the input amplitude. This is accomplished by the ADC producing a signal that is approximately the midpoint of the input voltage limits of the converter. The input signal is compared to this, but because multiple successive comparisons will be made on the same signal voltage, a sample-and-hold process is used in which the converter measures the voltage and creates a brief steady voltage of that magnitude; it can do this without measuring the voltage. The comparator determines whether the sample voltage is greater or lower than the test voltage. On the basis of that determination, a higher or lower voltage is delivered, which results in progressive narrowing of the window and accurately estimating the signal voltage.

Regardless of the mechanism used for performance of the analog-to-digital conversion, samples are taken at a high frequency so that the digital conversion produces XY coordinates that produce a good representation of the original signal.

Historically, ADC performance was a potential problem in data acquisition, with too slow rates of conversion resulting in distortion of high-frequency signal. This included *aliasing*, a term that has two meanings. The one to which we refer is where there is distortion of the signal by sampling at a rate too slow for the signal. Fine points (literally and figuratively) of the data are missed by the slow sampling rate. This is a non-issue in modern neurophysiological equipment.

After the ADC, the digitized signal is delivered to the computer for digital signal processing.

Stimulator

Stimulators are used in neurology, especially for NCSs and EPs.

For visual EPs, the stimulus is visual, either a checkerboard pattern that alternates dark and light squares or a flash stimulus. This is generated on a flat screen, with appropriate dimensions to fill the visual field of interest. The resolution, color balance, and rapidity of pixel display have to be sufficient to support the rate of change in stimulus, and all modern displays meet this requirement.

For somatosensory EPs and NCSs, electrical stimulation is typically by placement of electrodes over the area of interest, overlying the nerve under study. The device that generates the pulses is typically a square wave generator. In modern digital systems, control over the stimulator is digital, but the stimulator is analog. This means that there is a pulse generated, usually square wave. The parameters of the pulse are multiple, but the ones of greatest interest are polarity, amplitude, duration, and frequency. Some of the other features have to do with shape, current limits, and a few others. We assume that for this purpose, the shape is *square wave*. Other wave types are used occasionally.

Stimulation requires a minimum of two electrodes. In order for current to flow through the body, there has to be a path. When a person is accidentally shocked by external machinery or even lightening, there are two points of conduction. In the case of accidental shock, one end of the path is obvious, and the other end is some other part of the body. If the person is perfectly insulated and no current can get out even by sparking, current will not flow through a person.

Terminology for electrodes can be confusing. The stimulation that we use is *cathodal*. For a device producing electric shock, the *cathode* is where the electrons come out and the *anode* is where the electrons return to the stimulator to complete the circuit. This seems counterintuitive because cations are positively charged and anions are negatively charged. But that is because of the terminology of the ions. An anion is so called because it is attracted to the anode of a voltage source. Because the anode is positive, the anion is negative, so it is attracted to it. Similarly for cations, they are positively charged and attracted to the cathode, the negative terminal. Note that there is some excellent published writing which has this backwards.

At rest, there is polarization of the nerve with the inside negative relative to the outside. With stimulation, the extracellular milieu becomes more negative under the cathode and more positive under the anode. This results in there now being more negative charge extracellularly under the cathode so what was transmembrane polarization with inside negative is now less negative and ultimately positive because the tissue more negative outside the axon. This depolarization activates voltage-dependent sodium channels, which then start an action potential that propagates downstream. Accompanying this is hyperpolarization beneath the anode. This sometimes results in blocking of action potential propagation past that electrode. This is *anodal block*.

Signal Processing

Signal processing in neurophysiologic equipment consists of calculations to perform a number of tasks, including the following:

- Using the point coordinates to create a smoothed curve representing the original analog data
- Filtering specific frequencies and specific frequency bands
- Artifact rejection
- Optimizing display parameters

Curve generation is for ease of view. With high sampling rate, dots on a computer screen would appear somewhat smooth, but not quite, so smoothing is used. Our evolved brains prefer smooth surfaces and are better able to perform visual analyses than if we were looking at a scatter of dots. Of course, we need to ensure that smoothing is not removing important clinical information, hence this effect is limited to nonphysiologic frequencies.

Digital filtering is performed by complex math on digital data that suppresses representation of specific frequencies. There are many types of digital filters with different characteristics. For example, one type (termed *causal*) for a specific point considers the value at that point and also data occurring in time prior to the point of interest. An alternative method (termed *noncausal*) considers data after the point of interest as well as past data. How does the filter consider future points? The data analysis is performed on data that were acquired shortly prior to the present time, so some data from after the time point of interest are considered. Some of the digital filters used in neurophysiological analysis use fast Fourier transform to derive the spectrum of frequencies, which then allows for de-emphasis of select frequencies and re-presentation of the response in the absence of unwanted frequency components of the signal.

Artifact rejection is discussed in Chapter 3. Briefly, the artifacts from movement or other electrical interference are reduced or eliminated during digital processing of the signal. For example, signals are suppressed if they are too large or have a frequency composition that is inappropriate for the intended response. This is not just filtering but also active suppression of signal judged to be artifact.

Display

Multiple methods are used to display data. Paper is almost never used, although for many decades it was used for EEG recording. Nowadays, digital EEG recording and storage are not only more environmentally friendly but also allow for data manipulation that would not be possible with paper records.

Digital displays used in neurophysiological equipment and neuroimaging are high-resolution monitors with a large range of colors, depths, and brightness.

After data are acquired, digitized, recorded, filtered, and analyzed, the computer display shows the distilled data presentation. In the case of EEG, this is not only the channels but often also spectral analysis.

Display optimization is an underappreciated facet of neurophysiologic signal presentation. The display will be very different depending on the type of data. For EEG, we need at least 16 channels presented in an epoch where we can visually see the frequency composition that is important for EEG interpretation. This is very different from EMG display, in which different time and amplitude scales are shown without the multiple traces of the EEG. The optimization is not just to ensure that we have the right information displayed but also to ensure that it is displayed in a format that is appropriate for human physiology. For example, we do not want to have screens flash so fast that the brain cannot give a visual assessment of individual traces on the fly. An example is watching free run of an EMG recording. We might set the display time base so that the traces are flying by at 5 or 10 per second. Much faster than that and our eyes and brains are not able to see each trace as individual. For example, many commercial videos are at 24 frames per second. We do not see the individual frames; we start to lose perception of individual frames at frequencies faster than half that speed.

Audio and Video Recording

Audio and video capture are usually not central to neurophysiological studies, except for monitoring during EEG, where we want to be able to see and hear if a patient has a clinical event. Video recording overnight is helped by the camera being sensitive to infrared light. The display is not color, but we can see a clinical event to correlate with the EEG recording.

Audio recording begins with analog capture, with a sound pressure sensor that moves in response to sound waves. Subsequently, there is sampling of the pressure on the sensor, and this is then captured in digital format with a specific frequency bandwidth and sampling frequency.

Video capture is performed by a camera that has optics which can be focused on the patient. There is usually remote control so that the details of the video frame and focus can be adjusted remotely. Some also allow for remote control of target direction if a patient moves in the room. The video is usually captured by a charge-coupled device. A series of lenses focus and magnify as needed the photons from the scene. This is projected onto the detector. The photons impacting the device produce voltage that is dependent on the intensity of the light. This happens in a broad two-dimensional array. This analog signal is then turned into a digital stream of a standard format, with each data point being encoded for color (chroma) and brightness (luminosity). Note that not only is this detector used for the digital video capture that we use but also the video device is able to export the video as an analog stream in one of the standard video formats. When we adjust color, hue, brilliance, and other parameters of video, we are doing so on the post-processed video and not during the recording process.

Specific Physiological Studies

It may be instructive to discuss how display approaches are used for several neurophysiologic studies.

EEG uses computer display and the usual view is of a number of lines of EEG, which are digital representations of the EEG activity recorded from electrodes. Each line does not represent the activity of one electrode but, rather, is a function of the comparison of the electrical activity of that electrode to the electrical activity in one or a summed number of reference electrodes. A sample EEG screen is shown in Figure 2.17. Two portions of the recording are shown—one at rest on the left and with hyperventilation on the right. This highlights the benefit of digital display so we can easily do side-by-side visualization.

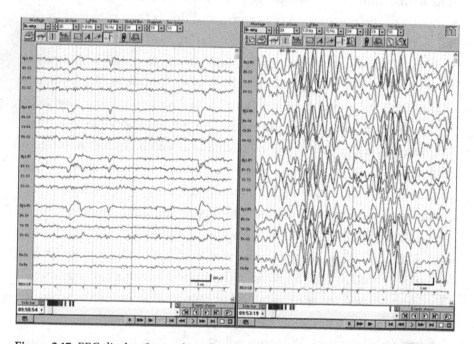

Figure 2.17 EEG display. Screenshot of an EEG display, with personally identifiable information removed. This is the EEG of a child. The left frame is normal awake baseline and the right is when the child is hyperventilating, which produces high-amplitude slowing.

Electronic display is used to show the results of NCSs and EMG. Because these two display formats are so different, there are effectively two screen displays that are used in EMG practice—one in which there is a graphic display of the nerve conditions on which measuring markers can be placed and one in which, during the needle portion of the study, a running screen similar to an oscilloscope display along with static images of selected potentials are recorded. Samples of EMG and NCS recordings are provided in Chapter 3.

EPs use a computer display that often has a running monitor of the potentials as they are acquired, and additionally has a progressing running averaged trace so the technician can determine whether the test is proceeding appropriately. Samples of EP recordings are provided in Chapter 4.

Computed tomography (CT) and magnetic resonance imaging (MRI) produce high-resolution images that are displayed on monitors which have specifications appropriate to the interpretation of images. Conventional computer monitors are not optimized for the task of displaying these images; higher resolution monitors are used. Samples of CT and MRI recordings are presented in Chapters 6 and 7.

ARTIFACTS AND ERRORS

A broad range of errors can occur during the course of neurophysiological study, including user errors as well as technical errors. Here, we discuss just a few of these errors.

Wire Failure

Most wire used for neurophysiological procedures is multistranded. As such, the wire is composed of a bundle of very small wires twisted together and then encased in plastic insulation. The multistranded character makes the wires more flexible than they would be if they were solid wire;

solid wires are very stiff and would be difficult to place on a human body. The problem is when one or more of the strands breaks. There may be no change in conductance if the other strands are intact and if there is little or no movement of the wire. But if multiple strands are affected, the resistance to current flow of the wire can increase dramatically. This is why all multistranded wire has a certain life expectancy—the strands will break at some point. Even if only one or a few strands are broken, there can be error introduced by movement of the wire causing make-and-break connections with the opposite broken end as well as other strands. A potential difference can develop across the two ends, and intermittent connection can introduce electrode pops and other high-frequency artifact. One take-home message is to not tightly coil wires, even phone charger wires.

Impedance Mismatch

All electrodes have an impedance, which is a resistance to flow. The term *impedance* is used because it means frequency-dependent resistance. Impedance mismatch occurs when the impedances of two or more electrodes are not equal. This results in a different amount of signal charge movement being picked up by the electrodes. Therefore, this can result in an error in signal recording.

Noise

Here, noise means electrical noise. Our environment is full of equipment, most of it characterized by a great deal of charge movement. As such, this produces electrical noise and also magnetic noise. The magnetic fields induced by charge movement then can affect the sensitive leads of neurophysiologic equipment, producing introduced errors.

Computers, Applications, and Artificial Intelligence

KARL E. MISULIS AND EAMON DOYLE ■

HISTORY OF COMPUTERS

Current computers are complex semiconductor-based machines with numerous connections internal and external to our institutions. The value of our machines is not just to effect and interface with our other technology but also to keep track of all of the knowledge because excellence in medical care depends on having the right information at the right time. We discuss the history briefly because tremendous advances within the past 30 years can suggest that all we use is modern technology, but the ideas of data storage, data representation, and assistance with tasks date back thousands of years.

The first data storage might be early writing. The earliest that we are sure about is Sumerian from approximately 3,000 BCE. There may be ancient writing in what is present-day Romania from 4,500 BCE, but the symbols have not been decoded and the authenticity is disputed. Nevertheless, data storage is ancient.

The first computing machines were mechanical and far predated the calculating engines of the 19th century. The abacus is likely the first computing device in widespread use. Earlier devices were more for counting and recording numbers and other data, and they were not value-added for computation other than storage of the variables.

The history of the abacus is best documented for the Old World, but a type may have been used in the New World as well. Types varied over time and geography. The abacus is still used in some locales.

Although there are different types of devices, an abacus of the type used in ancient China is shown in Figure 3.1. This discussion is relevant to the presentation of computer theory because analysis of the abacus can show how the sense of data representation, organization, and manipulation was developed long before modern computers.

Figure 3.1 shows the abacus in the home position with the upper sliders all up and the lower sliders all down. There are two sliders at the top and five at the bottom. As we look at the device, we see that it is organized as we would write numbers on a sheet of paper: The right-most column is for single digits, the next column is 10s, the next 100s, and so on. Interestingly, this has not been a universal consensus. Early Greek numbering sometimes had the least significant digit on the left, but this was not consistent even in that culture, and the Greeks' alphabetic representation of

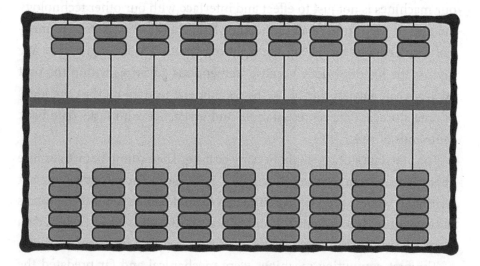

Figure 3.1 Abacus as used in ancient times. This one is base 10 and is only one form of the device. There are sliders on posts that can be slid up and down by hand. Waving the device back and forth while holding the central bar returns all sliders to home position.

numbers did not place the importance of string position that it does for our numbers.

The sliders in their present positions represent 0 for each digit. To make a 1, we take the right-most lower slider group and move the top of the lower sliders up to the crossbar. To make 2, we push another up, until we have all 5 of the lower sliders up at the bar. To add more, we with one motion move one of the upper sliders in that column down to the bar and push the lower sliders into their home position. So 5 lower sliders up or 1 upper slider down both represent 5, but the reset of the lower slides is to make the sliders ready to add more. Six is made by pushing one of the lower sliders up to the bar while one of the upper sliders is down on the bar. When we get to 10, all of the lower sliders are at the bar (+5) along with one of the upper sliders being down at the bar (+5), making 10. To reset this, the user simultaneously pulls the second upper slider down and pushes the lower sliders of that column into their home position. Two upper sliders down mean 10. To carry to the next spot, the user simultaneously pushes one of the lower sliders from the adjacent column up while pushing the upper sliders of the first column up to their home position. Although the device does not itself perform the math, the concept of using an abacus revolves around two critical pieces of knowledge required for computers: the states of the inputs and the rules that must be followed to produce a valid output.

One can see how addition and subtraction are facilitated with this mechanism. Its functions did not stop there, in that one could even perform multiplication and division.

Counting devices such as the one illustrated in Figure 3.1 use base 10, but construction could be adapted for different mathematical bases. Earlier counting machines used base 60, and there are examples of other bases used in other devices. It is easiest to illustrate the process with base 10 because that is what we use in our current day-to-day lives.

However, our advanced devices use different bases, depending on the function. Our computers understand our base 10 inputs but that is only because they convert all of our entries into binary, letters, numbers, punctuation, and commands. They also understand base 16 code (hexadecimal) that we use when programming our computers with the same translation. The hexadecimal system is a convenient method to bridge the native binary system with the human preference for numbers and letters in a word-type presentation.

Our ancient ancestors developed advanced analog devices long before the advent of modern computers. Through observation, complex mathematics, with trial and error, they developed maps and devices that together could predict the orbits of known celestial objects. This knowledge was used for navigation in prehistoric times, with constellations and asterisms (patterns of stars that do not form constellations) used long before any calculating device, but when the knowledge of motion and periodicity of celestial events was leveraged, remarkable assistance with navigation was noted.

One of the most impressive premodern devices is the Antikythera mechanism, probably built in approximately 100 BCE and found submerged at the site of a shipwreck near the Greek island of the same name. Extensive study of this mechanism seems to reveal that this device showed movement of the planets. This function was likely used for navigation on the seas. So although this usage cannot be decisively confirmed, the combination of this astronomical tracking function and the fact that it was found on the wreck of a ship support this theory. Some have termed this the first known analog computer, but we believe that the calculating devices such as the abacus and predecessors were, in fact, analog computers as well.

The Babbage difference engine of the early 1800s was the first mechanical computer. It was used punch cards for data and instructions.

The first digital electromechanical computers were electromechanical devices from the early-to-mid 20th century. Electromechanical devices are ones in which the data are stored as positions of relays, and the electronics of the device change the position of the switches.

Computers have been in widespread use in neurology as well as in many other specialties. Because of the nature of our specialty, our minds see a parallel between the analytic skills of the brain and the analytic skills of our devices. We are fascinated by the circuitry and capabilities of the brain, so we are interested in the mechanisms, complexities, and capabilities of computer technologies.

COMPUTER PHYSICS AND ENGINEERING

Computers are the modern black boxes; they do magical things but most of us do not have good understanding of what they do. Computers do math. At the core of a computer is the central processing unit (CPU), and attached to that are memory modules, devices, inputs, and outputs. Figure 3.2 shows a diagram of the CPU and some connections.

The CPU is composed of a control unit and a logic unit, the latter of which is often called the arithmetic logic unit (ALU). It is called this because it has two main basic functions. One is to do arithmetic, such as adding numbers. The other function is logic, such as determining whether one number is larger than another number or not. But these are both really arithmetic functions. Math is foundational whether the CPU

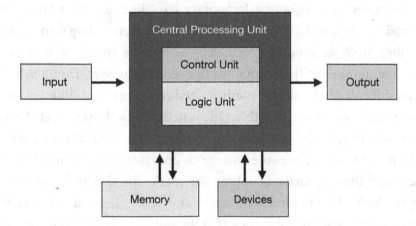

Figure 3.2 Central processing unit (CPU) with control unit (CU) and arithmetic logic unit (ALU). The CPU is made up of a CU, which keeps track of instructions, and an ALU, which performs them. Memory is used for operation as well as data storage. Inputs and outputs are in defined digital formats.

is tasked with calculating the product of two numbers or performing spell checking of a document or displaying electroencephalography (EEG) signal.

The control unit is the command portion of the processor. This keeps track of instructions and tells the ALU what to do. The instructions are written in a high-level computer language so the control unit is in part a binary decoder that turns the instructions into commands which the ALU can follow.

All calculations are performed in binary, where a bit has two potential values—0 or 1. Because multiple characters make multiple words, we use these bits to make bytes. A byte is a combination of bits to make a binary work of fixed length of 8. So proper byte examples are 10010110 or 00110110. These might stand for numbers or letters or commands. For example, 01000111 is uppercase G, and 01100111 is lowercase g. There is only a single bit difference. A carriage return is decimal 13, which is 00001101. A new line command is decimal 10, which in binary is 00001010.

Addition is performed by straight and computationally simple addition of the binary numbers. *Subtraction* is performed by placing the first number in a register, a temporary data site, and then taking the second number, making the negative of it, and then adding it to the first number. *Multiplication* is performed by adding the first number to a register the number of times equal to the second number. *Division* is the most difficult of the simple math procedures. There are different ways that processors can perform this. One is to do it exactly as we were taught to divide as children. Subtract the denominator from the numerator. Take the number of times the numerator goes into the denominator, and then store that, in binary of course. Then take the remainder, move the digital decimal point, and do it again and again. Then turn the binary number into base 10 and return that answer. This is the long way, and many processors do this. An alternative, more efficient way is to make a series of approximations, determine how far away from perfect they are, and get closer and closer until we achieve the degree of detail we need. Some processors use this method.

Complex functions such as spell checking involve the computer comparing a document to a library of words of the chosen language. If a word is not found, then the word is flagged for the human author, and a list of alternatives is presented. The list includes words that are spelled similarly to the entered word, where the difference is minor. With advanced spell checking, this is also often a word that sounds like the typed word, if spoken, because many people use voice recognition. Also, some spell checkers have a body of frequently confused words to look at and present as options in the case one is misspelled.

Analog-to-Digital Conversion

The calculations for EEG data presume that the data have been converted into digital format, and for this, an ADC is used. The ADC is an integral part of all neurophysiological equipment. ADC mechanistically was discussed in Chapter 2.

The analog signal recorded from the brain is amplified so that the signal is substantially larger than noise encountered during the acquisition process. Subsequently, the analog signal is converted to digital format by the ADC. The ADC samples the signal at specified times and makes measurements. Figure 2.16 illustrates the concept.

The rate of sampling is dependent on the hardware, but generally, most electrophysiological equipment samples fast enough to have a good representation of time-dependent changes in the biological signal. The sample is taken in a small fraction of a second; then the converter waits until it is time to sample again. The interval between samples is the intersample interval. The number of samples per second is the sampling rate. It is important that the signal be sampled at twice the maximum frequency present in the signal, known as the Nyquist rate, to be correctly represented.

In practice, the time resolution (sampling rate) and voltage resolution (amplitude of smallest voltage resolution levels) are so good that the reconstructed waveform would be virtually indistinguishable from the native waveform, and not a rough approximation as in Figure 2.16.

Signal Processing

Signal processing begins on analog signal prior to the ADC, but most of what concerns us as neurophysiologists is calculations performed on the digitized data.

Calculations can accomplish various tasks, including

- removing high frequencies;
- removing low frequencies;
- removing specific frequencies, such as 60-Hz line power artifact;
- determining amounts of specific frequency bands (power spectral analysis);
- identifying potentials that might be epileptiform activity (spike detection); and
- identifying sleep stages.

These calculations are not perfect. Frequency and timing distortion can occur, including phase shifts. Also, some functions, such as spike detection and sleep stage analysis, are particularly difficult for computers, so these functions are used for screening, and human review is essential for final determinations.

COMPUTER APPLICATIONS FOR NEUROSCIENCE

Neurology has some procedures, but mostly we are cognitive physicians. We interact with patients, families, health care providers, and others, and then depend on our electronic health record (EHR) and extensive diagnostic equipment that is highly computer dependent.

The EHR is discussed later because that is an element itself. First, we discuss the foundational principals of computer applications and the mechanisms they employ. Specifics of the computer science behind EEG and electromyography (EMG) are considered in the chapters on those modalities.

Data Acquisition

Data acquisition is recording physiologic information and depends on creating a connection between the patient and the data acquisition system. We first consider using electrodes to record electrical impulses from the body. In Chapter 2, we discussed elements of the data acquisition interface. Here, we combine some of these to highlight the complexity of this data flow from body to digital analysis system.

Figure 3.3 shows a diagram of the recording system for some sort of study requiring actual contact between the patient and the data acquisition system. This can be almost any study that uses two electrodes, EEG, evoked potential, or EMG. The top of Figure 3.3 shows the data transfer from body to the display, through all the individual units. Some of

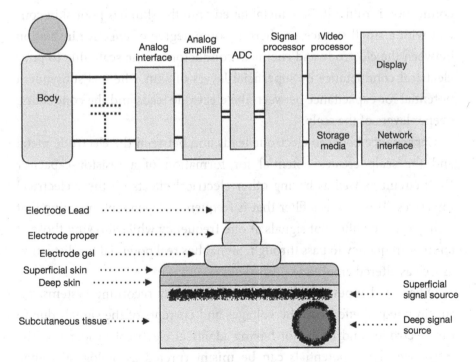

Figure 3.3 Complete recording system. Diagram of a recording system illustrating the complexity of the data transfer. (*Top*) The entire system from body to display. (*Bottom*) A close-up of the transfer of data from superficial and deep generators to the complex electrode–patient interface. ADC, analog-to-digital converter.

these are included in a singular box, but they are still separate. For example, ADC and amplifiers are totally separate physically but included
in one container, yet the interface complexity remains. The bottom of
Figure 3.3 is a close-up of the interfaces with data movement from
superficial and deep biological potential generators, as the potentials
are conducted through layers of skin, with some nearly devoid of
conducting ions, to the gel and the electrode, and then to the junction between electrode and lead wire, which is often a transition from
one metal to another. Any of these interfaces can be a source of error
or at least signal degradation. Here we discuss some key parts of this
system.

Electrode leads are just conductors, but their interaction consists of capacitance between the leads, especially if the leads are close together.

In the lead–head interface, the wire is attached to the skull, but direct
connection is difficult. Bare metal taped onto the skin has poor skin contact with a small surface, so there is some degree of effective insulation
between the electrode and chemical conductors of the scalp due to poor
electrical conductance by superficial layers of skin. There is subsequent
potential for capacitance between the electrode leads and the conducting
deeper layers of the scalp.

Capacitance between electrode leads and between the electrode metal
and the scalp creates potential for formation of a resistor–capacitor
(RC) circuit as well as having other electrical effects on these electrical
interfaces. This acts as a filter that is frequency-dependent, meaning that
it may resist or filter out signals at one frequency while allowing those at
another frequency to pass through. Signal has real potential for distortion
as well as altered amplitude.

Differential inputs, commonly used for many recording systems, result in unequal effects on the voltages and currents of the signals due to
the electrodes and leads not having identical electrical properties. The
introduced stray potentials can be misinterpreted as biological signal.
Also, variations in electrical properties of leads can distort the frequency
and amplitude of the recorded signal.

The bottom of Figure 3.3 shows one of the electrode leads attached to the scalp of the patient. The pathway from current from the brain to the electrode lead has multiple interfaces, each with potential to distort the response.

Another pivotal reason that the recorded signal is not a faithful direct representation of neuronal activity is that many of our recordings are by volume conduction. This is recording at a distance, like taking a photo with a camera that is out of focus. This has two effects. First, the geometry of the charge movement seen by the electrode is not the same as that of the generator because what we are seeing from a distance is a projection. Second, this is almost always the response of multiple neurons, almost always oriented in various directions. Therefore, what might seem cortically to be a fairly simple projection can result in a complex recording.

To illustrate the principles and potential pitfalls of volume conduction, we consider the difference between near-field potentials and far-field potentials.

Near-field potentials are where the generator is very close to the recording electrode. The electrode does not need to be in or directly on the generating neuronal membrane; in fact, it seldom is, especially in human studies. However, the electrode and the potential generator are in close proximity. Nerve conduction studies are near-field potentials. The electrodes are not in the nerves or even directly on the nerves; they are separated by skin and other subcutaneous tissues, but they are near.

Far-field potentials are where the generator is at a significant distance from the electrode. What constitutes significant distance is a fuzzy delineation, but we generally believe that when the generator is deeper than a small amount of tissue and especially when in a different biologic compartment, this is a far-field potential. Most somatosensory evoked potentials are far-field potentials. We often distinguish between near-field and far-field potentials on the basis of generation as well as recording geometry. Near-field potentials typically can be understood to be generated by a volley of activity in a nerve bundle, and we are recording the wave-front of that bundle. Far-field potentials are often produced by complex generators, often with divergent vectors, so even if you were to record close to the generator of a far-field potential, the recording would be complex. Last, one way to think of the difference between near-field and far-field potentials is that far-field potentials can be recorded over a much broader distribution compared to near-field potentials.

With some studies, both near- and far-field potentials can be recorded. Scalp-recorded EEG showing discharges of the cerebral tissue beneath the adjacent scalp and skull is near-field potential. However, when we see the other end of an EEG discharge dipole, that is far-field potential.

The purpose of this discussion is to illustrate a few points regarding the complex circuit that is a neurophysiologic recording system:

- Recording biologic signals is complex, with multiple opportunities for error.
- The physics and electronics of the electrode–skin interface are crucial, and poor attention to this one detail is a common cause of neurodiagnostic error. A small error in electrode placement, skin preparation, or any conductive gel application can degrade the effectiveness of the study.
- Often, the electrical recording is not at all the type of signal that the body is using. EEG waves are due to activity from millions of neurons, yet the function of the brain depends on the function of smaller groups of neurons. The far-field potential of a somatosensory evoked potential is a way of assessing integrity of the system. It is not a way to accurately measure function of the system.

Data acquisition is followed by data transformation. This is conversion of analog signals into digital data streams, as discussed previously. These days, everything is converted into digital format, even the colorimetric assessment of reaction vessels for analytic labs. The detector is analog, but the result is translated into digital format.

Data Analysis

Data analytics is a huge field, so we take a flyby of some of the methods that we depend on our data systems to do the analytics. Basic types of analytic processes include the following:

- Averaging
- Filtering
- Noise reduction
- Display optimization (smoothing, line drawing, etc.)
- Normal comparison
- Standard curve generation

Averaging is needed because there are often differences from trial to trial. If we were to consider all of the data, it would be an overwhelming task to determine which was the best single value to report from a number of Y values for every given X. Averaging smooths out the differences. Consider an X,Y data array. In this example, the X axis can be time from stimulus and Y can be the signal recorded. Because of subtle electrical differences and alterations in responsiveness from the subject, multiple Y values differ for time X.

The first form of averaging is to have multiple trials so there are multiple sample values for Y at each time X. We can average these to achieve an average Y. This is bound to be more accurate as we have increasingly more trials. However, there is a limit as to how many trials we can obtain. For evoked potentials, subjects tire of stimuli after a prolonged period of stimulation, especially if the stimulus is annoying or painful. Note that some electrical stimuli can be considered not painful on initial presentation but become so with repetitive delivery. There is a limit as to how many trials can be performed either due to patient tolerance or due to time considerations.

The second form of filtering is smoothing. We may have our own idea of what smoothing is, taking the bumps out of data, but sometimes the bumps are the areas of interest. So, there are a variety of algorithms, the simplest of which is *moving average*. The simplest form of moving average is to take the Y component of an XY measurement and replace the measured Y value by the average of the real Y plus the Y values associated with the adjacent X values. This assumes that a continuous recording is really incremental, with small increments, and this is always the case in neurodiagnostics. This is the 3-point average. When all of the points are considered equally (the index Y and the one prior and the one next), this is called *rectangular average*. The reason for this designation will become evident soon. If we believe that the measured Y value is more likely to predict the real Y for that X than the one before and after, then we may weight the Y values differently, so the real measured Y is perhaps given twice the weight of the Y values for the neighbor Xs. There is no reason to keep our smoothing isolated to 3 points. Why not do 5 or more? If we do, then we discount the contribution of the Ys associated with Xs more different from the indexed one. This is termed *triangular average*. The geometric terms rectangular and triangular describe the weighting of the adjacent values. If you place a data set that has been averaged next to a data set that has been smoothed, they will likely appear similar; they differ in that an averaged data set has had more data added to it that causes it to appear smoother, whereas a smoothed data set has had information discarded from it to make it smoother.

ARTIFACT REJECTION

A pitfall of neurodiagnostics is that we occasionally see large transients that are clearly in error, such as if a patient moves during a trial, causing the final recording to be significantly distorted. This is especially concerning because we are often making clinical decisions on the basis of small differences in values. Rather than relying solely on averaging, we leverage our analytics to perform *artifact rejection*. This is where a trial is considered so out of bounds that the associated data of that trial should not be in the final average.

We can consider two scenarios. One might be EMG or nerve conduction study (NCS) where there is one channel recorded and we have to decide whether a particular response is artifact on the basis of a single event. The other scenario would be EEG where we have other electrodes placed nearby the ones of concern, so we have more ability to determine what is artifact and what is not, although the math associated with this multi-electrode artifact rejection is quite complex. Whereas averaging is conceptually easy, EEG with multiple electrodes involves matrix calculations. Details are beyond the scope of this book, but we suspect readers fondly remember their matrix lectures in college.

Multiple types of abnormalities could be considered artifact by the recording system, including a response that is too large; one that is so small that it cannot be differentiated from unrelated activity; or one that has a frequency composition that is unexpected from the nervous system, even in the presence of abnormal function or disease. Another is when the timing is prior to some sort of stimulus, indicating that it was not due to the stimulus.

Analog filters are used for all neurophysiological equipment, but those are not the filters that we adjust. There are filters that determine which frequencies are passed to the neurodiagnostic device because ones higher and lower than a certain window are not going to be of interest. Historically, analog filters were the primary devices we would adjust, but that ceased long ago. The analog filters were circuits that would suppress frequencies higher or lower than a specified frequency. An RC circuit is a type of analog filter, although this was not used in practice intentionally. The RC circuit that mainly warped our recordings was unintended, with the resistor being the electrode leads and connection to the body, and the capacitor being the capacitance between electrode leads. This made a quite functional RC circuit that reduced response, especially to high frequencies. Thus, if a poor connection is made between a patient and the electrode, an additional RC filter is inadvertently introduced into the acquisition system. The operational analog filters use semiconductor circuits with properties designed to amplify only specified windows of frequencies and to relatively suppress others.

Currently, digital filters are the predominant filtering elements. These are algorithms with adjustable parameters that run on processors just like any other program. However, digital filters are in such common use that there are some application-specific integrated circuits (ASIC) which are purpose-built processors with configuration and programming specifically for digital filtering. Digital signal processors are similar but with a broader use, mainly in telecommunications, where they are adept at intake of analog signal. These specialized devices do not do any tasks that cannot be done by standard multipurpose processors, but their architecture allows them to do them faster. In discussing digital filters, we talk about the *transfer function*. This is a general term that means how one parameter varies in relation to another or, specifically, the transfer function describes how the output of a system is mathematically related to its input. This would not be just voltage in to voltage out. Rather, in the context of digital filters for neurophysiology, it would be the voltage output as a function of the frequency input. Certain frequency bands would be amplified to different extents or perhaps blocked altogether. This frequency-input versus voltage-output is the source of the transfer function term. The mathematics of these filtering systems gets complex, but we can conceptualize the abilities of these calculating systems to allow certain frequencies of interest and our abilities to change the parameters depending on need.

Fourier transform is introduced here. The specifics of the math are not important; most readers and certainly the authors are familiar with the basic principles but do not deal with the calculations. The concept is that the Fourier transform is a representation of a signal by determining the amount of each frequency in a signal. Not all signal is periodic, so this discussion is a little off-center, but a Fourier series is where a complex periodic function can be represented by the magnitude of each of a broad range of sinusoidal components. For example, a complex wave might be able to be broken down into the sum of a series of waves of different frequencies. In Figure 3.4, the darkest gray line appears complex, but it is the sum of a sine wave of 2 Hz of amplitude 3 summed with a sine wave of 6 Hz of amplitude 1. This is a Fourier series. The Fourier transform does this for our signals. Figure 3.4 shows a simple example of the Fourier

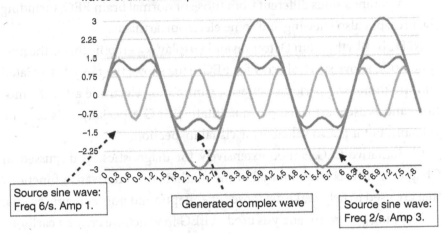

Fourier series: A complex wave is generated by 2 simple sine waves of different frequencies and amplitudes.

Source sine wave: Freq 6/s. Amp 1.

Generated complex wave

Source sine wave: Freq 2/s. Amp 3.

Figure 3.4 Fourier transform illustration. Breakdown of complex wave into composite frequencies and magnitudes. Fourier series in which the complex wave in dark gray is composed of two sine waves. The one in high-amplitude medium gray line is 2 Hz with amplitude 3. The low-amplitude light gray line is frequency of 6 Hz and amplitude 1. This illustrates how a complex wave can be broken down into fundamental frequencies.

series illustrating the Fourier transform. Complex signals can be broken into fundamental frequencies.

Power spectral analysis is related to Fourier transform and is the mathematical determination of the frequency composition of a signal. Whereas a Fourier series is a periodic pattern that can be represented by a series of sinusoids, and a Fourier transform is the process of transforming data in the time and amplitude domains into amplitude and frequency domains, power spectral analysis is a display of the frequency and amplitude during an epoch of a recording. This is most commonly used for EEG recording.

Spectral array displays the frequency and power over an epoch of time that is recorded. For EEG, this shows a frequency analysis for each part of the brain that a particular channel covers. This gives a good view of the frequencies that compose the signal—essentially a birds'-eye view of the power analysis. Usually, the spectral array is of an aggregate portion of the brain and then the user looks at individual channels to determine which ones are responsible for a particular presentation. Among the items that will show up on spectral array are those that would have frequency compositions different from those of normal brain EEG, including seizures, but also chewing and some electrode artifacts.

R2D2 or Rhythm Run Detection and Display is a graphic that is the next generation above spectral array for EEG. This shows the power associated with rhythmic and periodic patterns, eliminating electrode artifacts, motion, and muscle artifacts. For quantitative EEG especially, this is quite valuable, but it is also valuable for visual inspection.

Quantitative EEG is used extensively for diagnostics, as discussed in Chapter 6. This is a term for all of the mathematical derivative functions that are used to help make diagnoses of epileptic and non-epileptic events.

Mapping is a type of analysis used in EEG in which we create a cartoon-type appearance of the top of the brain, looking down at the scalp. Superimposed upon this are either colors or densities or isobars that relate to the topography of electrical discharges. These can be static or dynamic. This is a marked advance over just looking at the EEG trace, and it shows how digital analysis can clearly be value added.

Sound is not an analysis per se, but it is an extremely valuable way of displaying data in an unconventional method for EMG. Of course, we look at the trace on the computer screen as we are recording streaming electromyographic activity. However, we are also listening to it, and our ears along with our eyes are value added. So this is not a form of computer analysis but, rather, a different method of presenting data to facilitate our cerebral analysis.

Automated measurements are often used for neurophysiological studies, especially nerve conduction studies and evoked potentials. For NCS, a few single trials are obtained, and then a takeoff and peak are measured for many of the wave forms along with amplitude. The user typically adjusts these cursors, but the first pass at marking the parameters is often the province of the computer. This seems simple, but the algorithms used to make these measurements are quite complex, and there is an extensive body of literature examining different methods to make these as accurate as possible. In the case of NCS, there is ample time for the user to edit the computer's analysis. But with some applications, we need reliable algorithms that differentiate real from spurious recordings and correlate best with the physiologic process. Much of the literature in this field has been to validate the processes that have been developed.

Stimulators

Many neurophysiological studies require some sort of stimulation. This includes especially NCSs and evoked potentials. The stimulation is given by a signal generator. This generator is controlled by the computer system that is coordinated with the recording system. At a specific time, a specific stimulus with a certain frequency composition, wave shape, amplitude, and duration is administered. Acquisition is continuous; however, only data are recorded that are appropriate to the stimulation. For example, for an evoked potential, the recorder is always on but the only parts that are captured are those that are within a certain time window starting slightly before the stimulus and extending to a specified time after the stimulus. Recording slightly before the stimulus is important because if there is an artifact that will interfere with a recording, it is often visible prior to the stimulus. We know that that early epoch is not due to the stimulus.

Parameters of the stimulation have to be tightly controlled because responses frequently are dependent on these for consistency. This is especially the case because for every trial we record, hundreds and sometimes thousands of potentials are averaged.

Data Display

Despite advances in computer analysis, there is still no substitute for human analysis. This is especially true for neurophysiology and radiology. Computers may draw attention to certain areas of interest, but the human brain, with adequate training, makes the final decision.

Most of our studies are displayed on a computer screen, and some are also fed into audio systems. The methods and parameters are key. Knowledge of the science of our brains is essential to designing the best method of displaying these data. Factors that we consider include the following:

- Visual field
- Display parameters
- Highlighting
- Flicker fusion threshold
- Sound frequency response and amplitude

Visual field is an important aspect frequently underappreciated. Although we have clearly mappable visual fields, the part of the visual field that our brain pays the most attention to is foveal and parafoveal. Outside of these regions, our eyes will see data, but if there is something of interest our eyes will turn to look in that direction. This could allow us to potentially miss important data in another part of the display, which then becomes out of view. So, we can consider our visual field to be a target, with the greatest cognitive attention given to the center and progressively less as we get farther from the fovea. Therefore, we want a display that is going to allow us to see most of the data without moving our eyes, and we want most of the important data to be within the parafoveal region so that we will not miss the data. This has been studied scientifically, showing that people pay less attention to events and errors that are presented farther away from the fovea. Neurophysiologists need to keep this in mind when they are looking at the screen; if they get too close, they are making the problem worse, so backing up and getting the big picture is a good strategy.

Display parameters should be designed so that it is pleasing enough to the user to not be annoying yet to allow for optimal visual analysis. Brightness, color balance, and contrast are key, and with many devices, display parameters can be adjusted by the user.

Highlighting is where the computer system or a user has made measurements that are subsequently displayed with special prominence. For example, a technician may have highlighted a certain segment on a recording. This is extremely helpful but can be distracting from other findings that otherwise could be missed. Our eyes and brain are pulled to the area of highlight. So the highlights should be visible but not command our full attention.

Flicker fusion threshold comes into play when we are looking at streaming recordings, especially EMG. This is the frequency of display at which our brains are no longer able to distinguish individual pages from a continuous screen. The flicker fusion threshold for humans is approximately 16 Hz. Most video content that we see is at least 24 Hz, and often 30 Hz. That appears clearly continuous, without us perceiving the individual frames; however, for EMG, we need to see the individual frames, so less than 16 Hz is used.

We suspect that there is a purist in the audience that is of the opinion that we are misusing the word *Hertz*. Classically, the term is used to describe cycles per second of a wave; however, the term has been co-opted to indicate events per second, such as clock cycles of a computer. So in this sense, the usage is reasonable, although we appreciate a difference of opinion.

Data Storage

Data storage has become increasingly complex. In the past, data were recorded on paper, which meant that they had to be distilled data. Individual traces from a raster could not be recorded easily without quite some complexity to the graphics. The sound component was totally lost. Modern data storage is of two basic types, both computer based. Raw data are usually locally stored in the lab, whereas study reports are stored in the EHR.

Our enterprise computer systems are a compilation of multiple applications, interfaced so that they give at least some appearance of being a singular application. However, behind the scenes we have programs running on one set of servers for the EHR and others for associated applications such as radiology, clinical laboratory, and our neurophysiology data. The source is kept separately although interfaced.

Analog data are always recorded digitally. This requires that the sampling be adequate so that the reconstructed waveform is a faithful representative of the original physiological process. It is not uncommon for new analytic processes to become available, so data should be recorded for this future analysis.

There are specific requirements for how long certain data are kept. In practice, however, we almost never delete data. Even after a particular system is sunset, we maintain the source data indefinitely, usually in an archive form. The archive form may be on media not always online but can be retrieved if necessary.

Data storage is usually a combination of local and remote storage. Local storage is at the enterprise, whether it be hospital system or office. This includes the most recent system data. Remote data can be stored in multiple locations, either offsite in a different storage location, in case there is damage to the local systems, or it can be remote hosted, using a commercial entity such as Amazon Web Services or Oracle. These and other services not only have storage capacity but also have servers to run programs.

FOUNDATIONS OF THE ELECTRONIC HEALTH RECORD

The EHR is a key part of all practice. We may at times think of it as merely a method of entering data, reviewing data, and placing orders; however, the depth of complexity is almost invisible. Also, we are constantly trying to improve our informatics infrastructure to make the EHR not just storage retrieval and orders but also to help us take better care of patients, make fewer mistakes, and improve quality of care and outcomes. We have not yet met our expectations, and we suspect we never will, because our expectations will continue to move forward.

Here, we describe some of the essentials of the EHR, construction, capability, and future visions.

System Architecture

There are basic functions that are characteristic of the EHR. In addition, there are associated functions that at one point were optional but now are part of core functionality. Although the focus of this book is not on clinical informatics per se, from a neurophysics standpoint, it is appropriate to address how these functions support and influence our practice. The basic and associated functions are as follows:

- Basic functions
  o Data storage
  o Clinical documentation
  o Orders
  o Results
- Associated functions
  o Communication with patients
  o Communication with other providers
  o Decision support
  o Patient list management

The basic function of data storage was discussed previously and is similar for the EHR as our neurophysiological studies. However, EHR data storage is very centralized. The centralization process can occur in one or more locations, usually a combination of some local storage and remote storage. This remote storage may be operational or may be purely backup. Storage systems do occasionally fail, so redundancy is essential. These processes are essential to ensure data integrity. *Data integrity* is a specific term in clinical informatics that refers to the ability of the system to ensure that there is no data corruption, which could be introduction of errors or data loss. In other words, it is important that the system can store data, but it is also important that the system can identify when those data have been corrupted even if it cannot subsequently return the stored data. An associated function is *data security*, which is different from data integrity. Data security involves ensuring that no unauthorized personnel have access to the data while also ensuring that the data can reasonably be accessed by the people who need them. Security that is so tight that it cannot be reasonably accessed is counterproductive to providing quality health care.

Clinical documentation includes all of the data that we record as part of our health care work. Neurologically, this is typically associated with a substantial narrative component. Although some specialties, including urgent care, have very templated clinical documentation, this is not well-suited to neurologic care, in which fine points of history and examination would be lost. However, rating systems for strength and other physiological parameters are certainly valuable but are in addition to our narrative description.

Almost all orders are placed electronically. The spectrum of orders that neurologists place is broad, and the terminology may differ between ordering systems at different facilities and, sometimes, in the same enterprise, where more than one EHR program may be used. As a result, crucial elements of the orders component include use of synonyms and the ability to save favorites. In our enterprise, in which two major EHRs are used in different facilities, the names of the orders are often different. The ordering system needs to be flexible enough to allow clinicians to use multiple different systems and also to allow providers from other enterprises to join ours and not be hindered by a narrow lexicon.

Results are presented usually in one of two ways, either in list form or in tabular form, each of which can be displayed and sorted in several different formats. Graphic representation is important for select tests—for example, changes in creatinine during the course of an admission for encephalitis where nephrotoxicity from medications may be encountered. Doses of some medications, such as acyclovir, need to be adjusted based on these results. For radiology results, an embedded method of viewing images, sometimes called web PACS, is almost essential. Although these systems do not have all the robust characteristics of the fully functional PACS, they also do not have the overhead and can be launched from within the EHR.

Associated functions as listed are largely self-explanatory, but special comment is appropriate for clinical decision support, so we further discuss this later.

Database

Modern health care systems have multiple databases. At least one will be core for the EHR, and other databases are built for specific purposes. The following is a list of individual databases that an enterprise could employ:

- EHR core database
- Backup database
- Telemetry database
- Laboratory database
- Radiology database
- Data warehouse
- External reporting database
- Exchange database
- Research database
- Business database

The EHR database is core and the one on which the EHR works. There is usually a backup, which can be local to a unit or centralized. It is used mainly for data retrieval if the computer system is down. In some circumstances, network failure occurs, so only the backup database that is on a particular unit will function. This has a reduced data set, limited to particular patients, and sometimes only data from a single admission.

A telemetry database stores lots of information that we do not need. We only need a small subset of telemetry data as well as interpretive data placed into the core database.

Laboratory and radiology databases can be separate from core, especially if different vendors are used for these databases compared to the operational EHR database. An interface facilitates communication between these and the core database.

A data warehouse is where a subset of the EHR data is uploaded. This is usually used for analytics. Analytics routines can take significant resources from the search and retrieve services of the operational core database, so we try to run most analytics on a subset of information. For example, the data warehouse would have all results from the laboratory but not all of the point-of-care glucoses. Also, reports of radiologic studies would be in the warehouse, but the images would not.

The external reporting database includes data that keep track of infections, immunizations, and other items which state and federal agencies have determined need to be reported.

An exchange database is often used by enterprises to exchange data. Because most patients receive medical care at more than one enterprise, sharing data is key. Otherwise, patients are at risk of receiving inappropriate care.

Research and business databases are particular for those applications. Often, the research database has de-identified data, meaning that there are clinical data which can be analyzed but due diligence has been taken to remove personally identifiable information. De-identification is difficult because a modest amount of de-identified information can result in the ability to localize and potentially identify an individual patient.

Clinical Decision Support

Clinical decision support (CDS) is a major component of clinical informatics and is the career focus for many informatics professionals. Artificial intelligence will play an increasing role in this sphere, so this is considered separately. Here, we discuss the neurological implications of CDS.

A major reason why we need increasing decision support for neurology is because of the broadening expanse of treatment options. We discuss just a few of the conditions that we treat as neurologists: stroke, epilepsy, multiple sclerosis, and neurodegenerative disease. Decision support tools in these subspecialties are in various stages of development. For many of these conditions, there are so many possible treatments that humans could easily get anchored at giving their favorite treatment when it might not be the best one indicated for a specific clinical scenario. Relying on clinical data for decision support engines is not always as difficult as it might seem because the medical record will include reports from radiological studies, laboratory studies showing the status of metabolic systems, concurrent medications, and the host of rating scales that seem to be increasing almost daily, creating an acronym salad in the neurologic world. The rating scales are fairly standardized. Currently, radiological reports are not as standardized, but that is being improved so that if a patient has intracranial atherosclerotic disease (ICAD), for example, that fact will not get lost in the multiplicity of ways of saying so in a report.

STROKE

A major opportunity for CDS in neurology is identifying patients who are candidates for reperfusion therapy and also identifying which antithrombotic(s) is indicated for patients with select risk factors. Among the efforts in the stroke realm are decision support tools, some involving machine learning, to identify stroke suspects in triage.[1] Although this is not fully developed, it is hoped that the chance of a patient with an acute stroke sitting in the waiting room or exam room while they time out from the reperfusion time window will lessen. After stroke is suspected, there is a clear algorithm for inclusion–exclusion criteria for administration of thrombolytic therapy. This has been established using the foundation of the American Stroke Association/American Heart Association guidelines.[2] In most EHRs, these criteria are built as digital forms within the EHRs, and a particular agent is not to be used unless all of the inclusion and exclusion criteria have been addressed. Similarly, decision support algorithms are in various stages of development for guiding recommendations on post-stroke medical therapy, taking into account a variety of factors routinely available in the EHR in addition to rating scales, such as ABCD2 for risk of stroke after transient ischemic attack and CHADS2 for guiding need for anticoagulation for patients with atrial fibrillation.

EPILEPSY

Currently, 28 anti-epileptic medications are approved by the U.S. Food and Drug Administration. Keeping track of indications, contraindications, and drug interactions is extremely difficult. Although this effort is in its infancy, attempts are being made to improve this situation. EpiFinder is a decision support tool that is used for screening and diagnosis of patients with possible epilepsy; it is discussed further in the section on artificial intelligence.

Multiple Sclerosis

The number of agents available for disease-modifying therapy has expanded, which has made agent selection much more complicated. Humans tend to have only a limited palette of options for most problems. Just as there are many seizure medicines, there are many multiple sclerosis medications used for disease-modifying therapy, and although the indications overlap, assistance with selecting the appropriate medication is needed. The American Academy of Neurology has developed guidelines, but a robust decision support engine to help select the best agents is still lacking.[3] This would take into account numerous factors, most of which are already in the medical record.

Advanced CDS

Advanced CDS includes new techniques that can help diagnose specific disorders. One result of the Human Genome Project was the development of a type of spatial map of the human genome with associated information on function where possible. This was not performed on a single individual's genome, so the result is a compilation of data from many. Therefore, the goal was and is to determine the function of different genes and subsequently identify which disorders are associated with which gene mutations. The first target for this was select cancers. Now, this has been extended to other disorders. An example is the use of gene set enrichment analysis, which is designed to determine whether certain genes are associated with specific phenotypes. This has resulted in CDS support for diagnosis of select neurodegenerative diseases. There are clinical criteria for diagnosis of all of these disorders, but there is such clinical overlap that precise diagnosis on clinical grounds is often difficult, especially early in the course of the disease, when interventions to alter the course are most desirable. Also, not all patients exhibit all symptoms, and there is usually no specific combination of symptoms and signs that can be used across all patients. Lokeswari Venkataramana and colleagues developed the first generation of a CDS system that used analysis from patient data and gene set data, specifically for patients with Alzheimer disease or Parkinson disease. Accuracy of classification on the basis of gene products was approximately 80%. We anticipate that this will improve over time and similar models of gene and clinical findings will make diagnosis more accurate.

ARTIFICIAL INTELLIGENCE IN NEUROLOGY

Artificial intelligence (AI) is often discussed, but its role in medicine generally and neurology specifically is often not appreciated. We may think of AI as something on the horizon that may replace us as physicians. Rather, AI is a tool that will make us better. The analogy has been made by Penn's Kevin Johnson that AI is not making *Terminator*, it is making *Iron Man*.[4] It is not replacing the physician; rather, it is enveloping the physician in a very powerful tool to help with diagnostic and therapeutic decisions.

AI is not a promise for the future so much as it is an evolving reality now. Probably one of the most pervasive AI technologies currently being developed is autonomous driving vehicles. We do not expect an autonomous neurologist, but we do expect an increased penetrance of AI technologies. One of the principal tools of AI, machine learning, is increasingly leveraged to try to solve problems.

Machine learning uses a variety of algorithms to arrive at solutions. Rather than the programmer specifying the exact code set, the programmer provides a number of tools that the machine learning system can use to analyze a data set to draw conclusions. Then it builds models based on training data. Returning to the autonomous driving example, if you have ever "clicked all the images of a bus" to prove to a website that you are a human, you are participating in a large, distributed training data set to help teach computer vision systems how to identify things such as buses.

This is somewhat like giving the system an idea of what we want to accomplish. Then we turn the system loose on a broader set of data for which the answer is not known. Success or lack thereof results in modification of the machine learning approach until the best answer is received.

A noncomputer analogy might be a trained dog tracking a criminal suspect. The police officer need not know anything about how the dog accomplishes the task; however, the officer knows that the dog is induced to get a reward for finding the source of a scent to which it was previously exposed. Then, in an operational deployment, the dog is given a sample that smells something like a suspect, and the dog accomplishes the task without having to be taught how to do it.

EpiFinder is an AI approach to epilepsy diagnosis. This is a commercial venture partnered with academic epileptologists at Mayo Clinic.[5] The details of the engine are proprietary, but it uses clinical data, in addition to a repository of previous data with known answers, to predict whether or not a person has epilepsy and, if so, the type of epilepsy. Approximately half of patients evaluated in the epilepsy monitoring unit have non-epileptic events, either due to some biological process other than epilepsy or psychological in origin. EpiFinder predicts these diagnoses quite accurately. It does not have 100% sensitivity and specificity, but it is certainly a value-added tool.

Possibilities for future use in the sphere of cerebral activity include decoding neural activity into actual function. Joseph Makin and his team at the University of California, San Francisco recorded potentials intraoperatively using a dense array of electrodes in the perisylvian region while patients were speaking sentences. The recordings during each epoch were encoded and subsequently could be decoded into complete sentences. This was an almost perfectly accurate transcription from neuronal recordings, not from produced speech.[6]

Stroke is a target for some study using AI and machine learning technologies. Sirsat and colleagues reviewed efforts at applying machine learning to stroke diagnosis and treatment.[7] This review showed the expected finding that much more effort had been directed to stroke detection and characterization rather than treatment. They did identify some of the machine learning tools that were particularly able to assist with analysis of these data. This review highlights the research that is being done but also indicates an increasing need for machine learning assistance in treatment. As treatment options for stroke expand, decision support for the emergency department providers at stroke centers and especially at independent facilities is crucial to advancing stroke care. We have improving tools for diagnosis and treatment of acute stroke, but we need help ensuring that they are deployed and used appropriately.

Physics of Neurophysiology

EVAN M. JOHNSON AND KARL E. MISULIS ■

CONCEPTUAL FRAMEWORK

The nervous system is often described as the electrical wiring of the body. Although this is not terribly far off as a simile, there are considerable differences in how they conduct electrical signals.

In standard home wiring, electrons travel along a wire composed of a conducting metal, such as copper, insulated by a nonconducting material, such as polyethylene. The current of electrons (I) is determined by the equation

$$V = I \times R$$

where V is the voltage (volts) representing the potential difference between the anode and cathode propelling the electrons, and R is the resistance of the conductor given in Ohms. Much of this equation comes down to energy transformation: The electrons are compelled to travel from a state of high energy, where they are densely packed together and inherently repelled from each other's negative charge (like repels like), toward the lower energy state (opposites attract) in the bipolar configuration set up in the circuit. The resistance accounts for the imperfect nature of the conductor in which some energy is invariably lost across the length of the wire—less if a strong conductor such as silver or copper, more if a metal such as lead, and high if a nonconductor such as rubber. Advanced circuitry allows for complex on/off electrical conduction, but it pales to the complexity of the nervous system of humans and animals.

The conduction of electrical signals in living creatures is made possible by biophysics rather than a battery. The multipolar neuron engages in all-or-none transmission, decided by the level of input, both excitatory and inhibitory, it receives via its dendrites. If the input is sufficiently excitatory, it will set off an action potential transmitted through the axon to its end target by way of ion exchange along the length of the axon.

When inactive, the cell has a resting state cell membrane electrical potential created using a sodium potassium pump that continuously places three positive sodium ions outside of the semipermeable cell membrane in exchange for two potassium ions. The cell membrane does not permit charged particles to cross without the benefit of channels, so this results in a relatively high concentration of sodium ions outside and a high concentration of potassium ions within the cell. If all ions were able to freely intermix, cell membrane potential would be 0. Due to the uneven exchange, there are overall more positively charged ions outside the cell and a net negative charge within. Although it is named *resting potential*, it is the highest energy state of the environment.

To initiate a signal, sufficient excitatory activity must first occur. The dendrites receive excitatory and inhibitory signaling via neurochemical transmission. The activation of particular receptors by specific neurotransmitters may induce a more negative cell membrane potential (inhibitory) or a less negative potential (excitatory).

If the cell membrane potential reaches a certain threshold, voltage-gated sodium channels are activated. These channels open extremely briefly and then close. Although this is lightning quick, it is the perfect amount of time for charged particles to travel.

The sodium ions lined up along the phospholipid bilayer eagerly rush in at this opportunity. The end result of this influx of sodium ions is a shift in the local cell membrane potential, making it less negative. Although this will only affect a limited area near the voltage-gated sodium channel, it will very likely be enough to trigger the next nearest voltage-gated sodium channel, thereby setting off a domino-like chain reaction and sending this electrical signal down the axon to its terminus.

In the final phase of nerve transmission, the action potential triggers the release of a chemical signal. As the action potential arrives to the axon terminus, the shift in membrane potential will trigger a different voltage-gated ion channel, typically calcium. Calcium induces vesicle fusion with the membrane and ultimate release of the neurotransmitter from the vesicle to enter the synapse, where the neurotransmitter can activate associated receptors.

The nervous system has its own form of insulation. Schwann cells in the periphery and oligodendrocytes in the brain and spinal cord wrap the axons of certain neurons with several layers of a hydrophobic lipid-dense material, the myelin sheath. This ensures a more controlled environment along the axon, preventing ions from migrating away and ensuring peppy transmission speeds. Neurons without the advantage of myelin, such as pain and temperature fibers, are unlikely to send signals along faster than 10 m/sec. Myelinated fibers, such as the larger nerves controlling muscles, can transmit up to 12 times that speed. It is for this reason that demyelinating diseases in neurology such as Guillain–Barré syndrome or Charcot–Marie–Tooth disease can have a dramatic effect. More detail of the neuronal physiology follows.

PHYSICS OF EXCITABLE TISSUES

In this section, we discuss some details about the physics behind normal and abnormal neurologic function. First, we discuss normal functioning.

Neurophysiology depends on the physics and chemistry of excitable tissues. We learned the basics during our early training, but it is appropriate to revisit some of these principles in the context of our subsequent discussion of the physics of neurologic dysfunction.

Membrane Anatomy and Physiology

A brief review of the physiology of excitable tissues is appropriate. This is especially important because we are talking about the physics associated with the physiology and pathology we are dealing with. A simple diagram of membrane structure is shown in Figure 4.1.

What makes membranes excitable is that they have the ability to separate charge and to alter conductance through the membrane. These in combination allow for the electrical properties of nerve conduction, transmitter release, and response to transmitter action.

Membranes of excitable tissues are composed of lipid bilayers. For each layer, the hydrophilic ends of the chains are adjacent to the aqueous regions, intracellular and extracellular. The hydrophobic ends are in proximity to each other in the middle of the membrane. Proteins are inserted in and through the membranes, and these serve a variety of functions, including ion channels, a release site for transmitter, and receptor sites for transmitters.

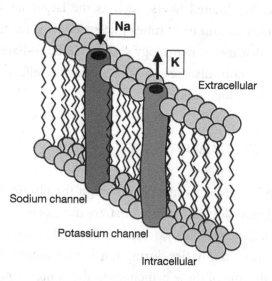

Figure 4.1 Membrane structure and physiology. Membranes are lipid bilayers that are both superstructure and a mechanism for compartmentalization, with transmembrane proteins providing the machinery for ion separation and gating.

Membrane Potentials

Polarization of the membrane is dependent on selective permeability or conductance (not exactly the same but related) to specific ions. The ionic gradient favors flow of sodium into the cell and potassium out of the cell. The channels have limited conductance, so the amount of charge movement is low. The sodium–potassium pump expels sodium from the cell and brings potassium into the cell, sustaining the resting membrane potential despite the small leak of ions down their chemical gradients.

Ions are affected not only by their own chemical gradients but also by electrical gradients. This is a difference in charge across the cell membrane, which is commonly approximately 75 mV in human neuronal tissue, with the interior negative compared with the exterior, hence termed –75 mV. Therefore, when channels open regardless of charge, there is a chemical drive to push sodium into the cell, and there is a chemical gradient to push potassium outside of the cell, on concentration consideration alone. However, the intracellular negativity is a further drive to have sodium enter the cell, and the intracellular negativity is a drive to have potassium enter the cell also, even though there is a concentration-dependent chemical gradient to exit the cell. So the directions of the chemical and electrical gradient affecting sodium are in the same direction—inward. This is a very strong electrochemical gradient. In contrast, electrical and chemical gradients for potassium are in opposite directions, so the specific value of the membrane potential affects not only the amount but also which direction the net electrochemical gradient for potassium is pointing. If the cell is depolarized, the electrochemical gradient for potassium is clearly outward. If the cell is hyperpolarized, the electrochemical gradient can be inward.

Equilibrium Potentials

The equilibrium potential is the potential at which all of the electrochemical gradients for ions that are able to pass through the membrane are precisely balanced. During this equilibrium, a small amount of sodium enters the cell, a small amount of potassium exits the cell, and the sodium–potassium pump uses energy in the form of adenosine triphosphate (ATP) to kick sodium out and pull potassium into the cell.

The membrane equilibrium potential is the potential generated by this charge separation and small amount of ion movement plus a small contribution by the sodium–potassium pump. This expels three sodium ions for every two potassium ions it brings into the cell, so this is responsible for –12 mV or less of the membrane potential.

The magnitude of the equilibrium potentials can be calculated from the concentrations of the ions. We first calculate the equilibrium potential for the major ions and then put them together with a weighting that is determined by the conductance of the membrane to each ion. This is because the membrane potential is mainly determined by the concentrations of the ions to which it is most permeable. The Nernst equation is used to calculate this equilibrium potential for ions assuming the membrane is only permeable to a specific ion (which is never the case).

The Nernst equation for potassium is as follows:

$$E_K = -58 \times \log \frac{[K]_i}{[K]_o}$$

where the bracketed element abbreviation (K for potassium, from the Latin *kallium*) is the concentration, and the subscripts i and o indicate inside and outside the cell, respectively.

Similarly for sodium (Na from the Latin *natrium*),

$$E_{Na} = -58 \times \log \frac{[Na]_i}{[Na]_o}$$

Similarly for chloride (Cl),

$$E_{Cl} = -58 \times \log \frac{[Cl]_o}{[Cl]_i}$$

Note that the inside and outside subscripts are reversed for chloride because of the negative charge of the chloride ion in comparison to the positively charged sodium and potassium ions.

We can calculate the equilibrium potentials for each of these. For potassium, the equilibrium potential is approximately −75 mV. This is the membrane potential at which the electrical gradient is equal to the chemical gradient so no potassium should flow. The potential is different for every ion.

To calculate the membrane potential, we take into account the conductances of each ion along with the concentrations. Again, positive and negative charged ions have flipped locations in the formula. This is the *Goldman equation*:

$$V = -58 \times \log \frac{G_K [K]_i + G_{Na} [Na]_i + G_{Cl} [Cl]_o + \ldots}{G_K [K]_o + G_{Na} [Na]_o + G_{Cl} [Cl]_i + \ldots}$$

where G is the conductance for each ion and the bracketed symbols are the concentrations inside and outside the cell. Again, the inside and outside subscripts for chloride are flipped because of the negative charge.

Using the Goldman equation, we can calculate the membrane potential. However, the important point is to understand how the membrane potential and conductance change in response to different values of the electrolyte levels. At rest, the conductance to potassium far exceeds that of sodium, so the membrane potential is closest to the equilibrium potential for potassium, whereas during an action potential, the conductance to sodium temporarily increases markedly, and that produces a membrane potential which is closer to that of sodium.

Changes in Membrane Potentials

There are several ways that a membrane potential can change. If one records from single neurons in the brain or spinal cord, there are many voltage fluctuations, and these are often not associated with action potential generation. This is because single neuron action potential often does not produce a single action potential in the central neuron system. It is true that a single action potential in a motor neuron results in an action potential in the innervated muscle fiber, but central neurons often do not have this property. Generation of action potentials is usually in response to multiple potentials delivered to the dendrites especially. And often, there is no action potential generation with every discharge. We believe that these spontaneous fluctuations are responsible for some of the complexity of data analytics and plasticity that our nervous systems exhibit.

Synaptic Transmission

Depolarization of a presynaptic terminal results in release of transmitter onto the postsynaptic membrane. Whether this is excitatory or inhibitory depends on which transmitter is used and the characteristics of the receptor.

Excitatory transmitters include acetylcholine and glutamate, and they produce their depolarization by opening sodium or calcium channels or both. The potential produced on the post-synaptic membrane is the excitatory postsynaptic potential.

Inhibitory transmitters include γ-aminobutyric acid (GABA), which opens potassium or chloride channels or both. The function of the inhibitory transmitter is not necessarily to hyperpolarize the membrane of the postsynaptic cell but more to clamp the membrane potential near the equilibrium potential for these ions, far from action potential threshold.

Transmitter that has been released into the synaptic junction could potentially activate the postsynaptic receptors again and again, but action of transmitter is instead terminated usually by one of three methods.

- Diffusion
- Reuptake
- Degradation

Diffusion occurs when there is a chemical gradient such that transmitter in the synaptic cleft diffuses away from the cleft and is no longer available. *Reuptake* occurs when the presynaptic terminal gobbles up the transmitter in the synaptic junction for reuse. *Degradation* occurs when the transmitter is metabolized to inactive substances. Acetylcholine and some of the neuropeptide transmitters are chemically degraded. GABA, glutamate, dopamine, and other catecholamines generally are removed by reuptake.

Action Potential

An action potential is produced when the depolarization of the postsynaptic membrane is such that a regenerative increase in sodium conductance is produced. That results in a marked change in polarity so that the inside of the neuron becomes positive with respect to the exterior, secondary to marked influx of sodium especially. Voltage-dependent sodium channels open resulting in significant depolarization which in turns opens more channels.

Action potentials are terminated by closure of the sodium channels in a time-dependent manner. In addition, there is opening of potassium and chloride channels, which helps normalize the membrane potential. Subsequently, the sodium and potassium that have moved between the intracellular and extracellular compartments are re-equilibrated by the sodium–potassium pump.

Potential Propagation

Potentials are propagated down neuronal membranes, but there is not always action potential generation. Depolarization of a focal region of membrane can produce electrotonic depolarization of cell membrane increasingly distant from the site of initial depolarization. Because there is no regenerative action potential generation, the potential will fall off, resulting in a lesser degree of depolarization. If the change in membrane was hyperpolarization, then certainly no action potential will be produced. However, much of signal movement in the central nervous system (CNS) is without action potentials because these subthreshold events still affect neuronal transmission of the cells, thereby affecting the downstream conduction of signal.

When an action potential has occurred in a cell, it can propagate by producing electronic depolarization of the membrane at some distance from the site of the action potential. This depolarization can be sufficient to cause an action potential at that distant site; propagation of this action potential down the membrane is the principal action potential conduction.

Depolarization of a segment of nerve that is myelinated (central or peripheral) results in faster conduction. The myelin sheath produces marked reduction in conductance of the membrane beneath the myelin, so only the nodes are available for generation of a new action potential. Because the transmembrane conductance is so much lower with the myelin sheath, the action potential essentially jumps from node to node, which is fast, efficient, and secure.

Muscle Physiology

Muscle is also an excitable tissue, capable of sustaining action potentials. The neuromuscular junction (NMJ) is very similar to a central synapse with some important differences. The transmitter is always acetylcholine, and no inhibitory transmitters are released at the NMJ.

Transmitter release produces opening of ion channels on the skeletal muscle, mainly sodium and calcium, which then produce depolarization of the muscle fiber membrane. For skeletal muscle, in the absence of disease, there is a muscle action potential for every motor nerve action potential.

Muscle fiber action potentials are propagated throughout the muscle fiber. The depolarization phase produces release of calcium from the sarcoplasmic reticulum, which then facilitates muscle contraction. Calcium and ATP facilitate cross-linking and release of actin and myosin filaments. The calcium is essential for the cyclic release.

Termination of the muscle contraction occurs when calcium reuptake halts the contraction cycle.

Feedback to the CNS regarding muscle contraction is controlled by several mechanisms. We classically consider muscle spindles in discussion of feedback. Muscle fibers that are responsible for the motor power are the extrafusal fibers, whereas the fibers responsible for loading of the muscle spindles are intrafusal. The terms derive from the fusiform nature of the muscle spindle. Because the intrafusal and extrafusal muscle fibers co-contract, a discrepancy between total muscle shortening and intended shortening results in activation of the muscle spindles, which then gives feedback to spinal and higher motor centers so that additional muscle activation can be produced.

Generation of brain activity especially with respect to electroencephalography and epilepsy is discussed in Chapter 6.

Neural Networks

The term *neural network* has two main uses. The one discussed here is a biologic network of neurons that function together to produce responses more complex than any single neuron could produce. The other type of neural network is a component of artificial intelligence (AI), in which the design is intended to be the computer correlate of a biologic neural network. However, the two concepts are only closely related by name. The AI form of neural network is not discussed here.

A biologic neural network cannot be described by a simple circuit diagram. It is attractive to envision a neural network as an electronic circuit with individual cells as power supplies and cellular membranes as resistors, capacitors, and transistors. However, what makes neural networks special is that connections are either enriched or suppressed based on the experience of the network. On a very simple scale, some neural connections exhibit *habituation*, in which repetitive activation results in lesser degrees of signal transmission, especially of neutral stimuli. The network focuses on transmission of signals with a particular priority, which might be translated into importance. This can happen even in the presence of significant noise. Hence, the neural network may detect and highlight the signal that otherwise would be lost in noise of the biologic system.

The flip side is *dishabituation*, in which the response to a stimulus that had been habituated is dishabituated when some separate stimulus enhances the response. An example often given on a macro level is when a person expects mail delivered to their office each day at a certain time and gives it almost no thought. If one day the mail is not delivered as expected, then the following day the person has heightened response to the mail delivery. A better example might be when someone has a heads up on a delivery of seminal importance, the response to the delivery is heightened. In either case, the habituation was dishabituated.

Habituation and dishabituation occur along with other paradigms in neural networks—repeated stimuli cause changes in connections and responses, also different complex patterns produce functional changes in the network. A multitude of related and unrelated events produce an almost infinite array of network changes.

The fluid nature of the neural networks makes the circuit diagram representation of the nervous system almost meaningless except for the most simple subsystems. This fluid nature is responsible for our neuronal excellence.

NEUROHYDRODYNAMICS

Fluid dynamics is the physics of fluids as we consider them here, but the field also includes movement of gases, which is not addressed in this book. The specific subfield of interest is hydrodynamics, which relates to the motion of liquids. For this discussion, we examine the classical hydrodynamic thesis of considering fluids to be a relatively homogeneous substance, termed continuous. This means that for purposes of the discussion, the complexity of blood and other body fluids with composition of cells and fluids is not considered, but other factors of the fluids are considered, especially density, viscosity, velocity, and pressure. Neurohydrodynamics is the specifics of fluid mechanics as it applies to neurologic systems.

Fluid dynamics impacts neurology in a few categorical ways. Intracranial pressure, blood vessels, and medical pump devices are all inherently tied to the properties of fluids, especially pressure.

In physics, many principles have to be segregated into ideal and non-ideal settings. Little in biology offers a perfect system. When considering the aspects of fluid pressure in the body, a key distinction must be taken into account: The skull is a relatively fixed volume. In most areas of the body, organs offer generous compliance, such as the bladder. This forgiveness is generally beneficial, even in the setting of pathological processes. In patients with liver failure, the human body can (begrudgingly) accommodate more than 5 L of ascites buildup in the peritoneum and more than 10 L in more extreme circumstances. The cranium; however, is not so accommodating, and the interplay between pressure and fluid dynamics can have a striking impact on brain function.

Intracranial Pressure

Intracranial pressure (ICP) is a product of the basic components within the skull: brain and meningeal tissue, blood and the vessels carrying it, and cerebrospinal fluid (CSF). There is some normal variance to ICP, usually in the range of 9–12 mmHg. An important relationship exists between ICP and the body's systemic blood pressure: cerebral perfusion pressure (CPP), the pressure maintaining adequate blood flow to the brain. CPP is calculated as mean arterial blood pressure (MAP) minus the ICP, where MAP is calculated as [(2 × diastolic pressure) + systolic pressure]/3:

$$MAP = \frac{(2 \times DBP) + SBP}{3}$$

where DBP is diastolic blood pressure, and SBP is systolic blood pressure. Thus,

$$CPP = MAP - ICP$$

Under typical circumstances, ICP is approximately 10 mmHg, and MAP is approximately 90 mmHg. This results in an ample CPP of 80 mmHg.

A few scenarios can threaten adequate CPP, resulting in impairment, syncope, or even death. Severe hypotension is one, as can be seen in shock, whether hemorrhagic, septic, or cardiogenic. MAP below 60 mmHg can threaten organ function. The other basic scenario is a pathological increase in ICP. ICP exceeding 20–25 mmHg is an indication for intervention. Because the skull is a fixed volume, situations of intracranial volume expansion can rapidly send ICP soaring. An example of this is traumatic intracranial hemorrhage. As pressure increases, it exerts compressive forces throughout the cranium on brain tissue and the walls of blood vessels. At critical levels, brain tissue can be irreversibly damaged, and the forward flow of arterial blood will be slowed and potentially arrested. The lack of circulation will kill additional brain tissue—ischemic injury. A third, less common scenario can also occur when jugular venous pressure (JVP) exceeds ICP. In this case, the venous system fails to pull blood from the brain back to the heart. This arrest of forward moving circulation also threatens CPP. In this scenario,

$$CPP = MAP \sim JVP$$

CSF is produced by ependymal cells along the inner lining of the ventricles and central canal. There is approximately 150 mL of CSF bathing the adult CNS at a given time, and this is being produced and reabsorbed at a rate of roughly 500 mL a day. Arachnoid villi covering the brain absorb CSF and deliver it to the venous sinus system.

An increased amount of CSF is an additional cause of increased ICP. Many conditions can injure and impair the arachnoid villi, such as scarring from meningitis. Our understanding of the mechanisms that underpin regulation of CSF production is still limited and under investigation and debate. In pathological states, any feedback system normally in place is either lost or insufficient; ependymal cells will continue to produce spinal fluid even if the CSF is not sufficiently absorbed. This brings about non-obstructive hydrocephalus, a condition in which excess CSF builds up. This causes expansion of the ventricles and will cause compression of the nervous tissue along with increasing the ICP. A lumbar puncture can help diagnose this condition as an abnormally high opening pressure is seen, and drainage of a large volume (30–40 mL) can provide temporary symptomatic relief. This condition must be identified quickly and a pressure relief valve put in place, typically a shunt. Shunts can be placed into a lateral ventricle and drain into the peritoneum, where fluid can be safely resorbed. Note that shunts and drains have the potential to inadvertently produce low-pressure headache and should also be checked for this in symptomatic patients.

If there is a non-obstructive hydrocephalus, it stands to reason that there is also an obstructive hydrocephalus. In this, the intraventricular path that CSF follows to the spinal canal is blocked or so narrowed that insufficient fluid can pass through. This can be caused by a mass lesion such as a tumor, a congenital structural lesion, or a number of other possibilities. In this situation, a lumbar puncture may not be helpful because an intracranial obstruction may lead to normal or low fluid pressure in the lumbar column while there is pathologically high pressure proximal to it. Head imaging can be valuable to identify the point of obstruction and inform potential intervention options. Typically, ventricles proximal to the blockage will expand. In some cases, a specialized imaging procedure called cisternogram may be necessary. In this, a radioisotope-labeled tracer (indium-111 tagged pentetic acid, as an example) is injected into the CSF and is serially tracked with a gamma camera. The tracer is commonly injected into the CSF via lumbar puncture, but it can also be placed directly into a lateral ventricle if the situation requires it. Magnetic resonance imaging (MRI) techniques exist that evaluate CSF flow without the use of injected tracer.

Multiple circulation disorders can affect CSF. In acute care neurology, the most common are idiopathic intracranial hypotension, low-pressure headache, cerebral venous sinus thrombosis producing reduced CSF reabsorption, and CSF infection or subarachnoid hemorrhage also reducing CSF reabsorption as well as worsening CSF pressure due to an inflammatory response. The commonality of the increased CSF pressure scenarios is that there is reduction in arterial perfusion because of the elevated downstream pressure.

Low-pressure headache usually develops after lumbar puncture, although spontaneous leaks can develop in the cerebral or spinal dura. Low spinal fluid pressure results in sagging of the brain and meninges when upright in the sitting or standing position, which can produce severe pain, both cerebral and spinal.

Ischemic Stroke

Vascular neurology concerns itself with the plumbing of the brain: Leaks and clogs are devastating to cerebral tissue because the brain is the most sensitive organ to oxygen deprivation. Ischemic stroke, in its simplest definition, is irreversible brain damage caused by nontraumatic lack of oxygen relative to metabolic demand. Most commonly, this is due to an interruption of arterial blood supply.

Arteriovenous anatomy shows that arterial stenosis or occlusion produces loss of oxygen and nutrients to supplied brain tissue, with the expected manifestations of ischemic stroke. Because the arterial tree has connections that provide for alternative pathways, the deficit can be lessened or even abolished by collateral flow. The ability of collateral flow to adequately supply neuronal tissue depends on the richness of the vessels, degree of vascular pathology affecting those vessels, and timing of the resupply. In some cases, this collateral flow may spare a patient from what would otherwise have been a devastating cerebrovascular attack. For example, more than 30% of patients who have been diagnosed with a transient ischemic attack, sometimes ambiguously referred to as mini-stroke, have signs of acute ischemia on MRI, indicating that there was clinical recovery but incomplete neuronal recovery.

The most common causes of ischemic stroke are embolic clots that form elsewhere in the body and lodge within an artery delivering blood to the brain or a local thrombus resulting from long-term vessel disease and plaque buildup. In either case, the tissue downstream from the blockage will be at risk unless there is adequate collateral blood supply or the clot is resolved.

An important tool in stroke workup is carotid ultrasound assessment, which is discussed in detail in Chapter 9. Duplex imaging involves both standard ultrasound assessment of the blood vessel and Doppler imaging. Doppler imaging can offer critical information, not only by confirming at least marginal flow versus complete artery obstruction but also by quantifying the rate of flow through stenotic areas of the vessel. Bernoulli's principle helps inform this: If a continuous flow of fluid goes through a volume of reduced radius, the velocity through that volume must increase so that the same overall number of particles are able to pass in a given amount of time. Whereas computed tomography (CT) and MRI are subject to a margin of error, the use of peak systolic velocity in carotid duplex assessments is a reliable method to diagnose vessel stenosis of greater than 70–80%, a key criterion in determining if a patient will require vascular intervention.

A third major cause of ischemic stroke is *watershed ischemia*. If there is stenosis of a great vessel, such as a carotid artery with heavy plaque at the bifurcation of the external and internal branches, the person could be vulnerable to a watershed event should there be relatively low blood pressure. This has been described as "a kink in the water hose." With normal, robust blood pressure, there is sufficient pressure to push past the narrowing in the great vessel and continue to supply the vascular territory fed by it. If MAP lowers past a critical point, however, insufficient perfusion may result. The brain does benefit from a clever evolutionary redundancy in the form of the circle of Willis. When fully formed, the circle of Willis permits cross-circulation across all the great vessels feeding the cerebrum (bilateral carotids and vertebral–basilar). In many people, this system is not fully formed, and in the case of profound hypotension, there still may not be sufficient perfusion to all areas of the brain. The overlapping boundaries of the various vascular territories tend to be the most at risk for loss in a watershed event, akin to a series of lakes drying up during a drought.

Other etiologies can result in ischemic stroke as well. In the case of a venous sinus thrombosis, the forward circulation of the arteries feeding the tissue that should be drained by the blocked venous system can become stagnant, preventing continuous supply of nutrients and resulting in ischemic stroke.

In mitochondrial disorders, the necessary efficient cellular metabolism engine is compromised and can lead to cell death in times of high metabolic demand. Stroke in mitochondrial disorders is particularly interesting from a neurophysics standpoint. For example, in patients with mitochondrial myopathy, encephalopathy, lactic acidosis, and stroke-like episodes (MELAS), there is a predisposition to stroke because of predisposition to vascular risk factors such as cardiomyopathy, seizure activity that increases metabolic demand, coagulopathy resulting in microhemorrhages, blood–brain barrier breakdown, inability to respond to increased metabolic demand resulting in regional ischemia, biochemical dysfunction producing heightened oxidative damage, and ultimately multifocal ischemia as well as hemorrhages spanning vascular distributions.[1] This is an example of how a multimodal failure of biologic and physical systems can have a common root cause.

In patients undergoing a sickle cell crisis, the clumping sickle cells can block small blood vessels in an embolic manner, exacerbating the already partially compromised efficient delivery of oxygen.

Marked increased hematocrit can produce arterial and venous stroke. In some patients, the underlying cause of the polycythemia can also produce disordered platelet function predisposing to stroke, but in the context of this discussion, we are referring to the hemodynamic effects alone. Elevated blood viscosity produces increased resistance to flow, slower flow with more edge resistance (edge of the column of fluid against the vessel wall), and therefore reduced cerebral blood flow. This can be seen in patients with polycythemia.

Hemorrhagic Stroke and Conversion

At the other end of the stroke spectrum is hemorrhagic stroke, which is heavily dependent on fluid dynamics. Among the characteristics of fluids is pressure, and each vessel in the vascular tree has a sensitivity to a certain level of blood pressure. The pressure is most pulsatile in larger proximal arteries, and there is less of a pulse effect with progressively smaller arteries and arterioles and only a small amount in the veins. The compliance of these vessels allows them the capacity to tolerate these fluctuations in intravascular pressure, especially elevations. Increased pressure, both mean and peak levels, chronically and acutely can exceed tolerances, predisposing to cerebral hemorrhage.

The major risk factor for any stroke is chronic uncontrolled hypertension. Atherosclerosis is the underlying mechanism for much vascular disease. In people with long-standing elevated blood pressure, greater than 140 mmHg systolic on average, the compliance of arterial walls is slowly lost. Arteries and arterioles are made up of three tissue layers, including a middle muscular one called the tunica media. It is this dynamic layer of smooth muscle that gives arteries the bulk of their ability to expand and contract to accommodate fluctuating fluid volumes and maintain appropriate perfusion pressure. Over time, excessive internal pressure leads to remodeling of the tunica media, with fibrous scar tissue replacing muscle and resulting in a subsequent loss of compliance. As vessel walls become stiffer, their ability to accommodate critically high pressures diminishes, leading them to become vulnerable to rupture.

The most susceptible vessels are small perforating arteries that directly branch from much larger ones. Examples include the lacunar arteries branching from the proximal middle cerebral artery to feed the basal ganglia, the thalamic perforators arising from the proximal posterior cerebral artery, and the basilar perforators. In each of these cases, wispy arterial branches arise from more robust vessels. Corresponding with the difference in size, the smaller vessel branches have far less capacity to withstand high pressures, although they are directly affixed to vessels that may have sufficient capacity to do so. Coupled with a chronic loss of compliance as an indirect result of hypertension, these vessel are at risk to rupture in the setting of elevated pressure, causing an intraparenchymal hemorrhage. Any mechanism for acutely spiking blood pressure, including ingestion of substances such as methamphetamines or cocaine, can trigger this.

Small vessels are also the most at risk for hemorrhagic conversion of ischemic stroke. The brain is an organ that utilizes autoregulation to maintain adequate blood supply. As local tissue has increased or unmet metabolic demand, nearby arteries are triggered to dilate and facilitate increased flow to these areas, increasing delivery of nutrients and removal of waste. This system is not switched off in the setting of ischemic stroke. Arteries in the stroke bed are prompted to dilate, although flow has been interrupted and blood cannot be adequately delivered. This ultimately leads to maximum dilation of the vessels distal to the clot blocking blood flow. In addition, there can be a Cushing reflex in which inadequate cerebral perfusion prompts an increase in systemic blood pressure in an attempt to increase CPP. If the clot were to be abruptly resolved by any means (tissue plasminogen activator, mechanical retrieval, etc.), the sudden restoration of blood flow could result in high-pressure perfusion rushing into vessels already maximally dilated and therefore without additional stretch to accommodate the influx. This can then lead to rupture of these vessels and hemorrhagic conversion. The aforementioned perforating vessels are again particularly at risk for this.

Malignancy can be another underlying mechanism for hemorrhagic stroke. Cancerous cells are defined by genetic mutations that largely promote relentless replication. Tumors have incredible metabolic demand, and many mutations gear toward promotion of adequate nutrient delivery to maintain this. Angiogenesis is a common finding associated with tumors, but cancers also prompt changes to the local vasculature, making blood vessels leaky as a way to maximize nutrient delivery. The structural weakening of these blood vessels makes them vulnerable to rupture. The aggressive malignancies most strongly associated with cerebral hemorrhage are the primary CNS tumor glioblastoma multiforme and metastatic cancers including melanoma, renal cell carcinoma, choriocarcinoma, and thyroid cancers. When a patient with a medical history of any of these tumors is found to have an intraparenchymal hemorrhage in an area of the brain not commonly associated with hypertensive bleeds, this should be suspected and evaluated with MRI.

Cerebral Amyloid Angiopathy

Cerebral blood vessels become vulnerable to bleeding in cerebral amyloid angiopathy. In this, β-amyloid plaques chronically affix to and weaken local blood vessels, making them friable. Microhemorrhages and later larger intracranial bleeds are seen, even in the setting of relatively well-controlled blood pressure. Gradient echo MRI sequences may reveal a vast number of previously asymptomatic bleeding events in these patient.

Vascular Malformations

Arteriovenous fistulas and other vascular malformations are other potential causes of intracranial hemorrhages. In an arteriovenous fistula, a "short circuit" direct connection between an artery or arteriole has formed with the venous system, bypassing the capillary bed. This bypass results in excessive pressure delivery across the conduit because veins are generally designed for low-pressure fluid delivery as opposed to muscular arteries and can be at risk for rupture. Cavernomas are another form of vascular malformation that lack structural capacity for proper fluid-pressure dynamics and can be a ticking time bomb.

Aneurysm

Aneurysms are a well-known source of potential intracranial bleeding. In them, a chronic weakening of the vessel wall can ultimately lead to ballooning of an outpouching. Should this pouch sufficiently dilate, the walls of the aneurysm may be so stretched as to be vulnerable to rupture, which can lead to devastating hemorrhage. This can result from vessel injury in the setting of long-standing hypertension or other mechanisms. There are several inheritable conditions in which people are more at risk for this, typically due to connective tissue conditions such as Ehlers–Danlos syndrome, Marfan syndrome, or polycystic kidney disease. If recognized and deemed to be severe enough, certain intervention procedures can be employed, including stenting, coiling, and clipping.

Endovascular coiling was developed in the early 1990s. In this procedure, a catheter is passed through the vasculature to approach the aneurysm. There, a set of small thin spring-shaped platinum coils are released from the catheter into the aneurysm in order to promote clotting and sealing off of the aneurysm. This is also referred to as embolization. An endovascular sheath is often also employed along the portion of the artery with the aneurysm to reinforce its walls.

In clipping, an external approach is taken, and a small metal clip is affixed to the base of the aneurysm to prevent further blood entry and dilation.

Vasospasm

Vascular resistance is increased by cerebral vasospasm (CVS) whether due to subarachnoid hemorrhage (SAH) or reversible cerebral vasoconstriction syndrome (RCVS). The most common cause of death with SAH is CVS. The pathology of post-hemorrhage vasospasm is not completely understood, but it is thought to be due to effects of leaked blood directly on the exterior of the arteries. It is an oddity that although blood is essential to life, almost any situation in which blood has been removed from vasculature containment leads to it irritating local tissue, often compromising function. The blood vessels themselves also react to injury. There are several evolutionary mechanisms to mitigate damage in the setting of traumatic injury, such as the clotting cascade. Vessel wall injury is a key initiator of this, as endothelial injury directly results in the release of local clotting factors. In addition, the presence of blood byproducts exposed to vessels outside of the endothelium prompts vessel constriction. In the setting of SAH, the release of free blood may inadvertently result in vasospasm of nearby arteries, impacting local blood supply to that tissue. Added to the vasospasm of SAH is increased ICP; together, these serve to reduce cerebral circulation and cause ischemia.

Reversible Cerebral Vasoconstriction Syndrome

Reversible cerebral vasoconstriction syndrome is a rapid onset of vasospasm that presents with acute "thunderclap" headache (abrupt onset of debilitating severity) which can be mistaken for a symptom of SAH. Associated with the headache can be cortical signs, which are focal neurologic deficits suggestive of vasospasm location, whether motor and/ or sensory or language. Seizures can also occur. The precise pathophysiology of the vasospasm is not known, but the presentation is dramatic. In some cases, it is associated with the use of substances known to have the potential to trigger vasospasm, such as cocaine.

Intrathecal Pump

Several medical devices also involve fluid-pressure dynamics to deliver therapies. Several pumps are becoming a helpful option to certain patients, including subcutaneous insulin pumps. In neurology, intrathecal pumps offer a method to deliver medications such as baclofen directly to the CNS, bypassing the blood–brain barrier.

Enteral Administration of Medication—Duopa Pump

In Parkinson disease, certain patients with advanced disease may find themselves with a narrow therapeutic window, requiring very frequent dosing of dopaminergic medication to relieve their symptoms but also with a predication for supratherapeutic effects such as dyskinesia. An external pump system is available to deliver a gel-based levodopa formulation directly to the jejunum to permit a constant rate-controlled delivery of medication to help these patients deal with their narrow therapeutic window without excessive pill burden.

Diagnostic Imaging of Vasculature and Fluids

In separate chapters, we delve into several of the imaging modalities available to facilitate the diagnostic workup for the many clinical scenarios discussed here. This section provides a brief summary of some of them.

FLUOROSCOPY

Conventional angiograms make use of fluoroscopy to permit clear assessment of the blood vessels supplying the brain. Continuous dynamic X-ray imaging of the head and neck permits real-time observation as a catheter is directed to a particular artery of interest in which iodinated contrast is injected. The serial X-ray imaging allows visualization as the contrast flows through the downstream branches of the artery. Areas of stenosis, blockage, aneurysms, and other pathology can be identified this way. This is also the gold standard for confirming vasospasm-related conditions such as vasculitis or RCVS. Vasospasms can be distinguished from focal stenosis by the administration of a calcium channel blocker, which prompts local arteries to relax but has no effect on stenotic plaques.

COMPUTED TOMOGRAPHY

Computed tomography is the key go-to tool in medical facilities throughout the world. Its relatively quick generation of three-dimensional internal imaging permits rapid initial assessment of patients; this is especially valuable when patients are unable to give an account of their history.

Non-contrast CT provides a basic anatomical scan, with excellent bone and lung tissue contrast. Although the dense cortical tissue of the skull leads to some inherent limitations in image quality of the brain, CT of the head (CTH) provides essential information in the time-sensitive stroke assessment. There is sufficient soft tissue contrast to reveal both hemorrhage (slightly opaque relative to brain tissue) and hypodense regions that could represent a completed stroke, tumor, abscess, or edema. The thrombosed clot blocking a great vessel may also be visualized. This information is necessary for determining whether thrombolytic therapy should be administration.

The addition of iodinated contrast to CT imaging of the head and neck can provide general assessment of the arteries supplying brain tissue; as large as the carotids and as small as some of the distal branches of the cerebral arteries. This can help identify occlusions causing stroke, the origin of intracranial hemorrhage, and other pathology. The digital subtraction of a registered CTH from a repeat scan with the administration of a contrast agent may create a virtual visualization of the vasculature without additional structures.

A dynamic imaging procedure termed CT perfusion offers key information in the acute stroke setting. A large bolus of iodinated contrast is administered, and a timed series of rapidly acquired CT scans are obtained. Post-acquisition analysis can then quantify cerebral perfusion with multiple metrics, including mean transit time, cerebral blood volume, time to max, and cerebral blood flow. This information is given for each element of the scan (pixels/voxels). Using this information, we can distinguish normal tissue from fully infarcted tissue (core) and tissue at risk (penumbra). This latter category is most likely to be amenable to reperfusion therapy.

In certain patients, imaging of the spine can be complicated by restrictions that preclude MRI. In these cases, an X-ray or CT myelogram may be preferred. In it, an iodinated contrast agent is administered, usually via lumbar puncture, and X-ray or CT imaging of the spine is subsequently carried out. This can facilitate identification of spine pathology, such as cord injury, tumor, or cysts.

MAGNETIC RESONANCE IMAGING

The various sequences of MRI can be invaluable in neurology, and MRI is the gold standard in diagnosing stroke. This is discussed in detail in Chapter 11 but is presented in brief here:

- Anatomical scans of the head and neck can be generated through T_1 acquisition.
- Diffusion weighted imaging (DWI) and apparent diffusion coefficient (ADC) sequences are sensitive to infarcted tissue and provide information on the relative acuity of the stroke.
- T_2^* weighted imaging, used to create gradient echo sequences, identifies areas in which the magnet spin–spin relaxation time (T_2) is affected by local magnetic field irregularities—that is, identification of the presence of substances such as collections of deoxygenated hemoglobin that would be seen in hemorrhage.

Gadolinium can be administered to provide contrast enhancement of blood vessels and areas of blood vessel permeability. In contrast to the iodinated agent of CT, gadolinium is paramagnetic and affects the magnetic field around it. This affects the relaxation times of immediately surrounding tissue and results in a hyperintense signal in post-contrast administration T_1 anatomical scans.

Time of flight is a type of sequence in which MRI scanners can visualize vasculature without the use of contrast. The sequence instead is generated from detection of areas with flow. In normal anatomical scans, blood vessels are "seen" as dark elements as a product of an artifact known as flow voids. The constant motion of particles disallows the determination and localization of their relaxation times. This inherent artifact is then changed into an opportunity, as a virtual maximum intensity projection can be created so that the areas of flow void are highlighted and create an MR angiogram or MR venogram. It is important to note that this technique is, in turn, vulnerable to artifact itself. In areas of slowed or stagnant blood flow, the flow voids essential to creating time-of-flight mapping may be lost. This can lead to the erroneous suggestion of an obstructed vessel rather than slow flow, an important distinction to correctly determine when there is concern for a venous sinus thrombosis. Despite limitations, the time-of-flight sequence is a method to assess vasculature inside and outside the skull without using a contrast agent, a benefit for patients with renal insufficiency or allergy to the contrast agent.

ULTRASOUND

Ultrasound can be a very useful tool for imaging areas of the body containing fluid, both because of the relative homogeneity of bodily fluids and because of ultrasound's ability to monitor flow via Doppler. Details are discussed in Chapter 9.

Duplex imaging is one example, permitting structural assessment in addition to dynamic flow evaluation. Peak systolic velocity can be a key determinant in whether a stenotic artery should be intervened upon in patients at risk for stroke.

Transcranial ultrasound is also used in monitoring patients at risk for vasospasm. Transcranial Doppler of both the internal carotid and the major vessels of the circle of Willis is a procedure that helps identify likely vasospasms based on flow velocity of the arteries as well as the ratio of velocity between the artery and the internal carotid artery (Lindegaard ratio).

Ultrasound can also be a handy tool in difficult lumbar puncture procedures, enabling clear visualization of the passages to the spinal canal available within the lumbar column.

NUCLEAR IMAGING

A nuclear cisternogram is a way of observing spinal fluid flow by injecting a radioisotope-labeled tracer into the CSF, most commonly via lumbar puncture but also can be administered directly into the lateral ventricles and serially scanned via a gamma camera. This is most commonly a SPECT imaging technique, using In-111-labeled diethyletriamin-epentaacetic acid

(DTPA). A CT scan can be performed in combination with this to provide anatomical information.

PHYSICS OF NEUROLOGIC DYSFUNCTION

Neurologic dysfunction can affect neuron bodies, axons, and/or myelin sheath. Neuronal damage is often due to the following:

- Mechanical effects from trauma, stretch, edema, or mass effect
- Infarction
- Hemorrhage
- Inflammatory change due to infection or immune attack
- Edema
- Degenerative condition

Axonal damage is often due to the following:

- Trauma, especially sharp but also blunt if severe
- Degenerative condition
- Metabolic process
- Vascular disease

Myelin damage is often due to the following:

- Mechanical effects of extrinsic compression, stretch, impingement by nearby tissues, local or regional edema
- Inflammatory change due to immune disease or infection
- Degenerative condition

Details of disorders are outside of the scope of this book, but some illustrative examples are presented in the section titled "Clinical Correlations." Here, we discuss the pathophysics and physiology of these different types of damage.

Response to damage is multifaceted. The type of response depends on the type of damage.

Neuronal Damage

The principal feature of neurons that makes them useful is there electrical properties. If they cannot move and modify charge, they are nonfunctioning. The final common pathway to many, if not most, neurologic disorders is failure of function of the neurons. The following are common mechanisms of failure:

- Loss of oxygen nutrient supply—for example, stroke and hypoxic ischemic encephalopathy, in which lack of oxygen results in short-term cell electrical failure and longer term cell death.
- Excitotoxicity—for example, Huntington disease, in which overproduction of glutamate results in cell destruction.
- Mechanical injury—for example, trauma or compression, in which there is damage to neuronal membranes; loss of the normal transmembrane potentials; and influx of ions, which can cause destruction of the cell unless very short-term damage.
- Toxin exposure—for example, ethanol toxicity produces neuronal damage in a variety of mechanisms, but one involves effect on a cascade in which there is a reduction in brain-derived neurotrophic factor, which is important for cell survival, especially in the hippocampus and cortex. Note that the dose required to impair initial neurogenesis is less than that which impairs survival, likely contributing to the increased sensitivity to damage with fetal exposure.
- Expected cell death—cells have a limited life span even under ideal circumstances.
- Axonal damage—mechanical damage to the axon or injury to the myelin sheath disrupts axonal transmission and can then cause more proximal neuronal damage. This is discussed below.

Neuronal damage does not always result in cell death. We do have some capacity for neuronal repair, as long as the insult is limited in time and severity. Neuronal regeneration is limited but can occur. An old belief that the adult brain cannot make new neurons is incorrect, but the ability is limited.

Axonal Damage

Axonal damage from almost any cause is followed by regrowth of the axons. The regrowth in humans is imperfect, so after trauma or other insult, not every axon regrows to its target. For the muscles that do not have their motor axons regenerate and for sensory regions that do not have their afferent axons regenerate, axons from nearby regions that do supply motor and sensory function try to compensate for the loss of neuronal function. This results in sensory regions that are larger than they were before injury and also motor units that have more muscle fibers per motoneuron axon than they had originally. Reinnervation after denervation is not perfect, but the body does usually compensate partially for the loss.

Myelin Damage

Damage to the myelin sheath can have multiple causes. The most common that we see are central demyelination from multiple sclerosis (MS) and peripheral demyelination by acute inflammatory demyelinating polyneuropathy (AIDP) or chronic inflammatory demyelinating polyneuropathy (CIDP).

MS is characterized by multifocal CNS demyelination, which can affect brain, spinal cord, and/or optic nerves. Although there has been extensive research, the exact immunological etiology is unclear, but there is evidence of inflammatory damage, and it is thought that T cell attack is responsible for the damage. The cause for that change in T cell response is unknown and might be triggered by some sort of infection in susceptible individuals. Nevertheless, the key for our discussion is that damage to the CNS myelin results in failure of axonal conduction. With the myelin sheath damaged, the axons cannot suddenly take on the electrical properties of an unmyelinated nerve, so conduction is impaired. If the damage is mild, then saltatory conduction can continue, although there is slowing through damaged regions. When the damage is more severe, conduction fails altogether.

AIDP and CIDP are peripheral neuropathies with demyelinating changes, characterized by motor and sensory deficits that develop in a sub-acute manner. Most patients have a monophasic disease, AIDP. However, a persistent and relapsing form is also common, CIDP. The damage is immune-mediated, but the cause is not known. Often with AIDP, there is a preceding infection, hence the recurring theme of infection triggering immune attack in susceptible patients. The cause of the motor and sensory symptoms is damage to the peripheral myelin sheath from a focal inflammation. The unraveling of the myelin sheath results in failure of the saltatory conduction. Similar to MS, if a particular region is mildly affected, conduction can be slowed, but damage typically results in failure of transmission at the damaged sites.

Note that MS, AIDP, and CIDP are discussed with respect to myelin damage, but they all have axonal damage as well. However, the myelin involvement is primary.

Compressive neuropathy is common and usually reversible. In carpal tunnel syndrome, the median nerve is compressed as it enters the hand from the distal forearm. For many people, transient pressure palsies can result in brief periods of single nerve dysfunction. When brief, the dysfunction is due to focal depolarization at the site of compression, blocking saltatory conduction. If the compression is not so brief, there is development of extracellular fluid, then inflammation, then further compression of the nerve. At this point, reversal of the damage can still result in resolution of the symptoms. If prolonged and severe, then there is subsequent axonal damage from which recovery is difficult, prolonged, and incomplete because the axons have to regrow, if they can.

Muscle Damage

Muscle damage not due to denervation is usually caused by degeneration, such as in muscular dystrophy, or inflammation, such as from an inflammatory myopathy. Muscle fibers try to regenerate, but this is incomplete. Because the damage is often ongoing, the degeneration–regeneration process results in an increased variation in muscle fiber size and in formation of excess connective tissue, which can further limit muscle performance because connective tissue does not have the compliant mechanical properties of muscle fibers and also does not have the contractile capabilities.

Neuromuscular Damage

Neuromuscular transmission disorders are commonly due to immune attack, in which antibodies bind to the prejunctional or postjunctional membrane and impair the normal one-to-one transmission from motoneuron axon to muscle fiber. We consider myasthenia gravis and myasthenic syndrome. We also briefly discuss NMJ dysfunction due to nerve agents, especially because that was a research focus of one of the authors.

Myasthenia gravis is due to a reduction in the action of acetylcholine at the NMJ. Most often, this is from an autoantibody to the acetylcholine receptor that produces weakness, as noted in the section titled Clinical Correlations.

Lambert–Eaton myasthenic syndrome is due to autoantibodies that act on the presynaptic terminal, impairing release of acetylcholine. So the final pathway is impaired NMJ transmission.

Nerve agents are a class of substances that attack the NMJ. Although the name implies that they should be a generic class of substances that attack nerves, the term is specific for agents that attack acetylcholinesterase, an enzyme responsible for degrading acetylcholine. This results in overstimulation at the junctions, producing a syndrome termed cholinergic crisis. Depending on the specific agent, the symptoms can vary, but central action is prominent. This starts with salivation and miosis (pupil constriction). Subsequently, there is gastrointestinal distress, respiratory distress, and ultimately jerking of the muscles that initially appears my-oclonic and subsequently convulsive. This is an example of augmented function of a neurotransmitter to the point of toxicity.

Botulism is blockage of the neuromuscular junction by the biotoxin that is a product of the bacteria *Clostridium botulinum*. This is one of the most potent toxins in terms of causing symptoms and death. The toxin appears to be internalized into presynaptic receptors and then damages transmitter release, thereby producing paralysis. Unlike nerve agents that overstimulate, this results in blockage of stimulation. The action is mainly peripheral because the molecule is too large to traverse the blood–brain barrier.

CLINICAL CORRELATIONS

Trauma produces damage to neuronal tissues in multiple stages. Initial trauma produces deformation of the neuron and/or axon. This immediately interrupts the electrochemical status, resulting in depolarization of the membrane. This can produce depolarization to the point that there is initially repetitive action potential generation. This can produce activation of downstream neurons and can be responsible for acute muscle activation with trauma. Subsequently, there can be persistent depolarization so that the normal membrane potentials are not restored. This produces inability of the neuron to generate action potentials. Hence, there is activation followed by loss of function. Trauma often affects vascular structures so that there is ultimately loss of function of the nerves due to ischemia.

Ischemia is loss of circulation to the neuron and/or axon. The loss of oxygen results in a brief period of preserved function, leading rapidly to loss of the ability to sustain membrane potential. Maintenance of membrane potential and channel function is essential to neuronal function. Immediate loss of function is due to hypoxia, and loss of glucose supply follows. Subsequently, there are later effects of the ischemia that make the damage much worse. One important element is glutamate. Although glutamate is an important excitatory transmitter, in higher amounts it is an excitotoxin. Ischemia adversely impacts glutamate metabolism.[2]

Myasthenia gravis is due to antibodies to the acetylcholine receptor. This is an autoimmune condition, and in most instances there is no cause, although a minority of patients have a thymus tumor. The antibodies bind to the acetylcholine receptor, which prevents binding of acetylcholine. The antibodies also cause internalization and destruction of the acetylcholine receptor. The absence of a functioning NMJ results in weakness and fatigue. The weakness varies depending on activity because the synaptic security of the NMJ is incomplete.

Guillain–Barré syndrome (GBS) is a demyelinating neuropathy that most commonly presents with weakness and sensory deficit with subacute progression. GBS is a family of autoimmune disorders, the most typical of which is AIDP. Classically, there is a preceding infection to which antibodies are made. Subsequently, these antibodies cross-react with the myelin sheath of peripheral nerves. The exact targets depend on the type of antibodies. In the case of CIDP, the myelin sheath is the predominant target, and damage to it results in slowed conduction velocity on nerve conduction study and also failure of some axonal transmission, producing weakness.

Physics Behind
Neurologic Technology

PART 2

Physics Behind
Neurologic Technology

History of Technology in Neurology

KARL E. MISULIS AND EVAN M. JOHNSON ■

FROM GUESSWORK TO PRECISION CLINICAL DIAGNOSIS

It is human nature to want to do something to help people. When the current status of medicine does not have effective or available tools, we fall back on trying treatments that are unproved. Most of the time, these are ineffective, but occasionally the findings can be revolutionary. This rare justification to the process of guessing psychologically validates our ideas and approaches to treatment. We may believe we have come far from the times of trial and error, but that is where we are now, just that we are being a bit more scientific about it.

The scientific method has revolutionized our advancement of knowledge, beginning with asking a question, formulating a hypothesis, gathering data through either observation or experiment, and then revising the hypothesis accordingly.

So although we sometimes look down upon the earlier efforts at science, our predecessors discovered fascinating truths and efforted extraordinary accomplishments on the road to the present day. When we design a drug and then search through multiple disorders for a response, we are not always being so much smarter.

EARLY USE AND MISUSE OF TECHNOLOGY IN DIAGNOSIS AND TREATMENT

In a far off land a long time ago, a loved one fell ill or was injured and was on the brink of death. The people of the day were used to death, life was shorter and harder than for most of us today, yet there had to be an impetus to do something to relieve suffering or to reduce the risk of death. Most of these attempts were ineffective and actually harmful, but they were attempts nonetheless.

Here may be the appropriate time to quote Voltaire: "The art of medicine consists in amusing the patient, while nature cures the disease."

Trephination is the practice of drilling holes in the skull. It was done for uncertain reasons but presumably to release some substance responsible for disease. It is difficult to know exactly where and when the first use of trephination was employed, and there are many ancient skulls that have been found in which holes were placed at the time of or after death. But examination of skulls has identified skull defects that showed signs of remodeling, indicating that the patient survived for an extended period. Trephination is still performed for neurosurgical procedures, including drains for excess cerebrospinal fluid and blood and as an access point for brain biopsy.

Bloodletting is the most often remembered treatment that was ill-advised. The theory was that there were evil humors which were responsible for sickness, and by draining the humors responsible for the disease, one might improve. When a person or two would survive the procedure and their disease, the practitioner would take credit for the cure. Of course, bloodletting is still used for hemochromatosis, but the primitive indications were far beyond that.

Electrical stimulation was used for medicinal effects in ancient Egypt and Rome. The electrical activity of torpedo fish was known and considered to possibly have beneficial effects. This kind of electrical activity could activate muscles that were otherwise paralyzed, so this treatment was used for patients with stroke and other paralytic condition.[1] This treatment did not disappear with these ancient peoples; in fact, even in colonial times electrical stimulation was used for stroke. Ben Franklin investigated the role of stimulation in stroke recovery but did not find it effective. He did not advocate it for his wife as she was suffering from a series of strokes.[2]

RISE OF SCIENTIFIC TECHNOLOGY IN NEUROLOGY RESEARCH AND PRACTICE

The birth of scientific technology was careful testing of various technologies to determine whether they could give valuable information or perhaps be of benefit. This was the first step on the way to all of the technologies that we have today. We eventually learned to have control groups, comparing data from an experimental population against the control population. We learned to compare old versus new technologies to determine which accomplished our objectives better. The scientific methods of observation and experimentation had been born.

Electroencephalography

The first generation of electroencephalography (EEG) machines were preceded by recording of electrical activity of a variety of biological tissues. The first recordings of brain activity were probably from electrodes placed directly on the surgically exposed brain. When taking an exam with a question regarding who invented the EEG, one should answer Hans Berger, but that is actually the wrong answer. Forty-nine years prior to that discovery, Richard Caton, a physician working at the Liverpool Royal Infirmary School of Medicine, recorded the electricity from living and dead brains of animals. He did these recordings directly from brain tissue as well as through the scalp. He found that brain activity increased during sleep and then eventually disappeared with death.

Subsequently in 1924, Hans Berger was the first to record electrical activity from the human brain. He recorded and described different waves, including what we now call the alpha and beta waves. His reported electrical activity from humans was recorded from the scalp. This was not accepted by the scientific community initially, likely because it was known that the skull produced marked attenuation of the electrical activity of the brain, and it was easier to record directly from the cortical surface.

EEG was initially recorded on paper, and only relatively recently was the transition to electronic recording and storage made standard. Since those early days, the field has advanced with increased numbers of recorded channels and ultimately migration from analog paper records to digital electronic records. This migration was not only an advance in display technology but also provided for digital analysis, which has become a science in itself.

Electromyography

Electromyography (EMG) dates back hundreds of years, although not used as it is today. It was known in ancient times that electricity could be produced by chemical and physiological mechanisms. Recording of the electrical activity from voluntary activation was probably first recorded by Nobel Prize winners Joseph Erlanger and Herbert Spencer Gasser, who viewed the electrical potentials on an oscilloscope of the day.

Both EMG and EEG have benefitted greatly from the miniaturization of electronics, the invention and utilization of integrated circuits, and the partnering of modern computers with neurophysiological recording and stimulation equipment. While we are seeing the same potentials as those seen in the past, our ability to display and store has markedly improved usability.

Intraoperative Monitoring

Intraoperative monitoring has become commonplace in advanced surgical suites, especially for brain and spine work. The development of this technology required equipment to be sufficiently small yet sensitive and compliant to place in a surgical suite. Electrocorticography, recording directly from the cortex, is often performed prior to resective surgery for refractory epilepsy. Somatosensory evoked potential monitoring is often performed during spine surgery in which conduction is at risk. Brainstem auditory evoked potential monitoring is sometimes performed when surgery in the area of the cerebellopontine angle is required.

Critical Care Monitoring

Continuous EEG monitoring is done in critical care units for an increasing number of indications. Over time, we have realized that we missed cases of nonconvulsive status epilepticus (NCSE), and because this can have irreversible implications for neuronal function, we often have EEG placed in critical care units when NCSE is considered a possibility. Also, patients can go in and out of electrical seizures, so we are more likely to perform long-term recordings, at least 24 hours at a time.

Common scenarios for performing critical care EEG monitoring include the following:

- Known status epilepticus
- Monitoring for frequency and recurrence of seizures
- Suspected nonconvulsive status epilepticus
- Post-cardiopulmonary resuscitation (CPR) encephalopathy

For patients who are post-CPR, continuous EEG is often performed. This is to ensure that seizures do not develop, especially as sedatives are withdrawn.

TECHNOLOGY WITH UNCERTAIN IMPLICATIONS: THERAPEUTIC HYPOTHERMIA

Not all technologies stand the test of time, and this is also part of the scientific method. We continuously evaluate data and revise our decisions based on new data and analyses.

Therapeutic hypothermia for patients status post cardiac arrest has been used for the past several years. We had known from previous anecdotes that patients who were hypothermic tended to be more likely to have better neurologic outcomes after cardiac arrest. We have examples from our own hospital in which patients were resuscitated when they were profoundly hypothermic and had better than expected outcomes. Therefore, studies were performed that examined lowering of temperature to 32° or 33°C. The optimal magnitude of the required hypothermia was uncertain. The optimal duration of the induced hypothermia was also uncertain.[3]

The theory behind therapeutic hypothermia is that there is a reduction in cellular metabolism, which would theoretically reduce demand when stressed. The reduction in adenosine triphosphate production due to hypoxia results in a number of changes, including influx of ions through the cell membranes. This can then activate enzymes, which can damage the cells further. When cooled, the influx of these ions is reduced.

The damage is exacerbated when oxygenation is restored. Oxidative stress during reperfusion adds to injury, with an associated inflammatory response. Studies were performed to examine whether protocolized targeted temperature management (TTM) might be effective. Data seemed to show that there was benefit to cooling to 33°C.[4,5] Subsequent studies indicated that there was not a significant difference between this lower temperature and 36°C.[6]

A more recent study, TTM2, indicated that there was no benefit of therapeutic hypothermia over merely enforcing normothermia.[7] These are very complex studies, with comorbid conditions and differences in critical care, and these fine points are likely responsible for the difference in results.

The conclusion for this discussion is that theoretically therapeutic hypothermia should work. In practice, even under tightly protocolled performance, there is some ambiguity, which basically means that we do not know all of the variables and we are not controlling for them adequately. Among the issues that we might focus on are ensuring no post-rewarming hyperthermia, as well as hemodynamic and sedative protocols.

From a neurophysics standpoint, our point is that this type of technology stands on observations, supported by some controlled study, contradicted by some controlled study, and clearly involves so many variables that definitive determination of efficacy and the exact protocol that should be used is so far elusive. This means that we need to be even better about our technology, get a better handle on all of the variables, to firmly determine whether this intervention is effective. It is an important issue, because this and many other interventions are very resource-intensive, so we want to ensure that we are gaining from them.

This general issue applies to many of our technologies. We have to balance the effort, expense, and complexity against the outcomes. In the absence of unlimited resources, we need to choose our weapons as well as our battles.

Electroencephalography

KARL E. MISULIS AND EVAN M. JOHNSON ∎

BASICS OF ELECTROENCEPHALOGRAPHY PHYSICS

The discussion of electroencephalography (EEG) physics assumes a basic understanding of cellular biology and particularly the physiology of excitable tissue, including neurons. This text focuses more on physics rather than physiology. However, we discuss physiology as it applies to EEG generation. We first consider generation of normal EEG activity and then briefly review the physiology behind some abnormal EEG activity.

Neuronal tissues have a resting membrane potential that fluctuates on the basis of connections. Most of the changes do not result in action potentials; a multitude of neuronal inputs are needed to bring a neuron to generation of an action potential. Some of the synapses are on the neuronal dendrites and some are on the cell body or soma. In general, synapses on the dendritic spines are more likely to be excitatory, whereas synapses on the soma and dendritic shafts are more likely to be inhibitory.

Scalp EEG is recording from many neurons. Although it is attractive to consider EEG spikes and sharp waves to originate from single action potentials, that is not the case. Scalp potentials require the fairly synchronous activation of many neurons. These potentials are volume conducted through the skull and scalp in order to be picked up by scalp electrodes.

GENERATION OF EEG ACTIVITY

Normal EEG Activity

Normal EEG activity consists of a wide range of activities. We discuss only a few of these here. Fundamental to this discussion are the biophysical mechanisms of how the electrical activity we record as EEG is generated. It is attractive to think of the EEG activity as being produced by action potentials of individual neurons or perhaps groups of neurons, but we have found that action potentials are not the major contributor to the charge flow that we see. Summed postsynaptic potentials are the major contributor, both excitatory and inhibitory.

Consider the cerebral cortex, which underlies most of the scalp. The cortical ribbon contains large neurons that project to other cortical regions and to subcortical structures. Pyramidal tract cells are just one minority segment of this population. For every one of these neurons, there are many more that synapse on the dendrites and body, and the charge movement associated with these connections is thought to be responsible for most of the charge movement of scalp potentials. Summed activity of many neurons is required. One estimate is that summed activity of 6 cm² of cortical surface is required to produce what we see on the scalp.

The implication of this arrangement is that activity of many neurons is required for scalp EEG to be generated and abnormalities to be detected; abnormalities in small numbers of neurons can be invisible. Another implication is that some parts of the cortex do not project activity onto scalp electrodes. For example, gyrations of the cortex are such that some neurons within the sulci are much deeper to the scalp electrodes, and their dendritic arborizations are not oriented perpendicularly to the scalp so that activity in those neurons will not be seen in the same amplitude and manner because of these difference in space and projection. Even more disturbingly, there is extensive cortical tissue that does not underlie the scalp, such as interhemispheric, inferior frontal lobe, and medial and inferior temporal lobe. These are regions that commonly are the focus of seizures, especially the temporal lobe, yet they are almost invisible to scalp electrodes unless the discharge spreads far beyond the region. To detect seizures originating in these regions, special techniques are required, including placing electrodes inside the skull or inserted them into tissues below the base of the skull, which cannot be done with scalp electrodes.

Normal activity has a wide variety of patterns, but to illustrate EEG generation, we consider only two—the posterior dominant rhythm (PDR) and sleep spindles.

PDR is an approximately 10-Hz wave seen over the posterior skull regions, overlying the occipital cortex. Figure 6.1 shows an EEG with a normal PDR. The PDR is sometimes considered to be an idling rhythm of the brain, although the metaphor with engines is far off target. This is sometimes also called the alpha rhythm because in the normal adult waking state with eyes closed, the rhythmic activity is in the band of EEG frequencies termed *alpha*, which is 8–13 Hz. However, it is not a very useful term because if the waves are not in the alpha range (faster or slower) the label would no longer apply. The generator for the PDR is thought to be an interaction between the occipital cortex and thalamic pacemakers. This is just one of many rhythms where there is cyclic relay of information between cortical and subcortical centers, part of the continuous neuronal activity and connections in the brain.

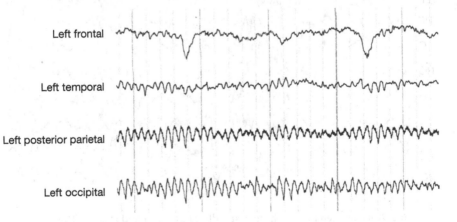

Figure 6.1 Normal waking EEG, with a normal posterior dominant rhythm. EEG of an adult in the waking state with eyes closed. Each major vertical marker is 1 sec. There is an approximately 10/sec posterior rhythm, which is normal.

Sleep spindles are another rhythmic activity, but whereas the PDR disappears in sleep and is replaced by other frequencies, sleep spindles appear in the central cortical regions during light sleep and disappear with deeper sleep and with awakening. Figure 6.2 shows sleep spindles in a child. These also are thought to represent cyclic activity in connections between cortex and thalamus. The role is not completely known, but it is thought to play a role in memory consolidation, in which short-term memory is integrated with long-term memory.

The generation of the PDR and sleep spindles does illustrate an interesting conceptual observation. Our brains are often compared to computers, but of course there are many differences. One that applies here is tasking. A computer is asked to do a task and it does it, and then it waits for the next task to be assigned. Brains are fundamentally different. When not doing something else, such as looking at an image, the neurons have no significant reduction in activity but, rather, are engaged in a host of internal associative tasks that we believe are responsible for much of

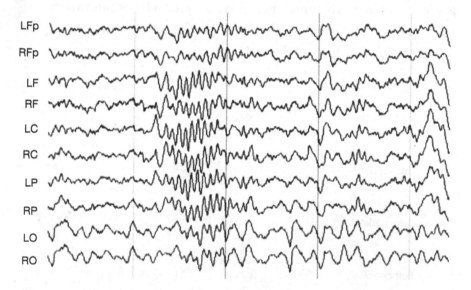

Figure 6.2 Normal sleep EEG with sleep spindles. EEG of a child at sleep. The background activity is a mixture of frequencies that indicates sleep, being different from the previous waking record. Left of center is a high-frequency rhythm, the sleep spindle, which is present at light sleep and disappears with deep sleep.

the special features of brain computing—creativity, association, subconscious processing, revisionist memory, and others.

Abnormal EEG Activity

Abnormal EEG activity has a broad range of manifestations. We select just a few EEG abnormalities to illustrate how they are generated and why they are abnormal.

Spikes and sharp waves are discharges that correlate with seizures. Samples are shown in Figure 6.3. Spikes and sharp waves have similar implications but are differentiated by the sharpness of the wave. Sharp waves have a longer duration, whereas spikes have a shorter duration. These are scalp manifestations of the synchronous activation of a large section of cortex. We noted previously that for many waves to be detected on the scalp, at least 6 cm² has to be involved, but for spikes and sharp waves, studies show that usually more than 10 cm² is involved and often 20 cm² or more. Spikes and sharp waves not only are scalp manifestations of the discharge of many neurons but also represent the repetitive discharge of most, where an episode of depolarization results in a burst of potentials from each.

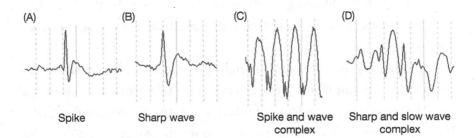

| (A) | (B) | (C) | (D) |
| Spike | Sharp wave | Spike and wave complex | Sharp and slow wave complex |

Figure 6.3 Spikes and sharp waves. EEG signals representative of some abnormal discharges. (A) A single spike as would be seen between seizures. (B) A sharp wave, with a similar clinical implication but it has a longer rise time and duration. (C) A spike and wave complex as would be seen with a seizure. (D) A sharp and slow wave complex, which also can be seen in patients with epilepsy.

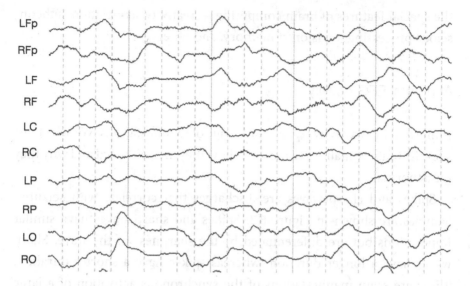

Figure 6.4 Slowing. EEG of an adult with severe hepatic encephalopathy. Marked slowing is evident, and there is disorganization of the background.

Slowing is when the scalp potentials have more low-frequency activity than would be expected for the age and state of the patient. Figure 6.4 shows severe slowing from metabolic encephalopathy. A large amount of slow activity is expected in deep sleep, but that type of slowing in a patient who is not asleep is abnormal and suggests encephalopathy with a broad differential diagnosis. Focal slowing is almost always abnormal and indicates structural damage near the area of the electrodes picking up the slow activity. Slowing is believed to be generated by the disruption of the pathways between the cortex and subcortical structures, especially the thalamus.

The definition of abnormal is slightly complex and a dynamic discussion. One might think that determination of normal and abnormal could be established by calculating electrical activities expected with different pathologies; however, it is rather more clinical correlation—what activities are seen with different pathologies. In that way, interpretation of EEG is more complex than knowing the electrical connections of the brain, and it is not possible for us to comprehend these connections in-depth.

EEG ACQUISITION AND ANALYSIS

Electrodes are placed in standard positions to cover most of the scalp that overlies cerebral cortex, as shown in Figure 6.5. The positions are standardized so that we do not need to look at novel specifics of placement on the patient to interpret the findings.

The electrodes are made of a material that does not promote polarization and thereby allows for the signal seen by the electrode to be a good representation of the underlying cortical activity. The electrode leads connect to a junction box, often called a head box because it is connected to the head. The box has sockets for the terminal ends of the electrodes—a design that facilitates correct placement of the electrodes on the head.

The head box usually contains amplifiers that are responsible for boosting the signal from the electrode leads to a higher voltage. This results in lower susceptibility to noise as the signal is then sent through the box's cable to the EEG machine. Without this amplification, there

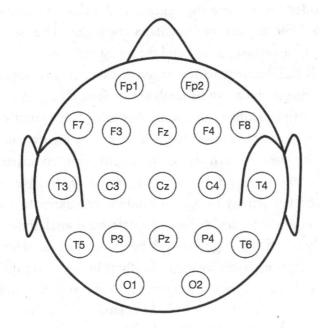

Figure 6.5 Head diagram of electrode positions. The abbreviations are standard, with F, T, P, O, and C standing for frontal, temporal, parietal, occipital, and central, respectively.

would be potential for greater electrical interference from the environment. Hospitals are equipment heavy, and all the resultant current movement produces magnetic fields. These magnetic fields produce current movement in electrode leads and other mission critical neurophysiological connections through the process of induction. For the cables from the head box to the EEG machines, the voltage and current have to be greater in order for the environmental electrical noise to not be problematic.

Acquisition then includes relay of the information to the computer systems after digitization.

Display

Data arriving at the EEG device include not only the recording of brain activity but also physiological activity, which is helpful for interpreting EEG. This often includes eye movements and electrocardiogram (ECG) and sometimes respiratory movements. These are recorded using surface electrodes for the eye movements and ECG and transducer for respirations. Note that eye movements can be detected because the eye is a dipole with the cornea positive and the retina negative.

Display is standardized with a raster of channels that together show a complex image which can be analyzed in seconds by an experienced reader. The displays differ somewhat depending on manufacturer, but they have an appearance much like that shown in Figure 6.6. Each line represents the electrical activity of one electrode in comparison to the electrical activity coming from one or more electrodes. We say "more" because often the activity in one electrode is compared to a calculated average of other electrodes. Because the data are digital, the calculations are relatively trivial, although they can be made more complex if certain electrodes are given greater weight than others in the aggregate reference.

The inset photo in Figure 6.6 is a running video of the subject being recorded. As we acquire and view the EEG either in real time or later, the video is synchronized with the EEG displayed so we can see if there is a particular behavior that correlates with the EEG findings.

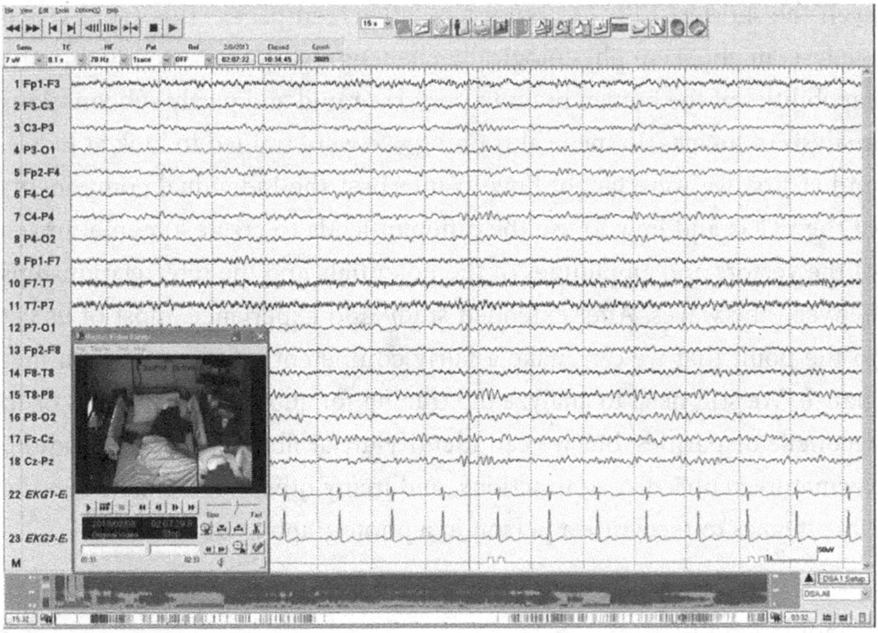

Figure 6.6 EEG display. EEG display that includes an inset image of the patient so that any behavioral correlate to EEG findings can be seen. This is a real-time display, but the video image is also available for display on the stored recording.

Analysis

We look for abnormalities in rhythms and for electrical discharges that might be seizure activity. By knowing the topographic correlation of the display with the head, we can then determine the location as well as the character of the discharges. We can change the topographic representation of the lines on the display to highlight different parts of the brain, but we do try to keep fairly standard maps or montages so that our own brains learn rapid visual analysis.

When clinicians are new to looking at EEGs, they are often amazed as how quickly an experienced electroencephalographer can view the EEG pages, reviewing the traces much faster than acquisition speed. This is a skill that is learned from looking at hundreds or thousands of EEGs, and because we usually view only 10 seconds per displayed page, even the shortest EEG (2 hours) has 120 screen pages. This rapid visual analysis might seem physiologically challenging, but it is well within the capabilities of the human brain to perform rapid and partly subconscious analysis of images. In medical training, we were trained to look at ECGs, and at first we were taught how to interpret the individual components of the ECGs and how to use the different leads to create a mental image of the vectors and amplitudes of the potentials and the interrelationships between the waves. After extended study and experience, most of us get to the point that we can make a fairly competent interpretation on first glance. We still need to analyze the tracing for subtle abnormalities or to reconcile disparities, but we can identify atrial fibrillation, ST elevation, premature ventricular contractions, and many other abnormalities in the time it takes to recognize a person in a photograph as our friend.

Quantitative EEG

Digital EEG allows for detection methods that were not possible with historic analog EEG recording. Quantitative EEG (QEEG) is an umbrella term for the advanced data handling that allows for several functions, including the following:

- Seizure detection
- Spike detection
- Mapping
- Spectral array

Seizure detection is a function in which data are analyzed and the combination of electrocerebral and noncerebral activities suggests that the patient is having a seizure. This might be association of body movement with rhythmic activity in the brain. Alternatively, it might be synchronous discharges from connected cerebral regions, with a coherence that would not be expected with normal resting EEG activity.

Spike detection differs from seizure detection in that a behavioral event is not detected. Rather, there is the occurrence of an electrical event in one or more channels that the onboard algorithms judge to be possibly a spike. Although the detectors are far from as discerning as a human, they can bring our attention to potentials that deserve further consideration.

Mapping is a function in which a diagram of the brain has colors and intensities that correlate with frequency components and amplitudes of activity.

Density spectral array (DSA) is a calculation performed on the frequencies and produces a color-coded visual representation of the power of different frequencies in the EEG. Changes in DSA can indicate seizure activity or state change (wake, sleep). In Figure 6.6, the gray-scale line near the bottom of the display is the DSA for this patient.

These QEEG techniques will undoubtedly become refined with more clinical experience, but considering the multiplicity and complexity of the variables, we suspect artificial intelligence technologies will be involved in the next-generation QEEG techniques.

CLINICAL CORRELATION

Figure 6.7 shows EEG samples from patients with the clinical correlations discussed here.

Normal EEG

The spectrum of what is "normal" is broad. There are differences in recordings depending on age, wake–sleep state, level of alertness, and in relation to physiologic status, such as eyes open or closed. The waking and sleep records are rich with a spectrum of activity. In general, activity is faster when awake, slower with sleep, and slower still with deeper stages of sleep. During the waking state, there is a posterior dominant rhythm, sometimes considered an idling rhythm, considering the penchant for humans to form an analogy between objects, including brains and their cars. This idling rhythm is approximately 10 Hz.

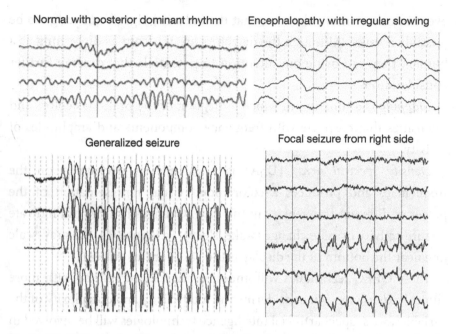

Figure 6.7 EEG samples. Normal, encephalopathy, focal-onset seizure, and generalized seizure are shown as labeled on the individual frames.

Encephalopathy

Activity is slower than normal, with the posterior dominant rhythm more slow and less regular, showing more variation between waves of the rhythm. Slowing and reduced rhythmicity is due to disordered relays between cortical and subcortical structures, so there is less synchrony and more variability in the response timing.

Generalized Seizure

The normal EEG is replaced by high-voltage synchronous discharges, with a spike component and a slow-wave component. The rhythmic nature is usually due to circuits between cortex and subcortical structures. The abnormal electrical activity results in loss of the activity needed for consciousness, and the prominent rhythmic discharges usually result in rhythmic motor activity, often clonic or jerking. However, the motor manifestations can be tonic muscle activity without and clonic phase, and with absence epilepsy, there can be little or no motor activity. Discharges are high voltage and sharp because of the synchronous activity of many neurons, entrained as a group. Although it is attractive to think of these as summed action potentials, it is rather summed field potentials from the discharge of multiple neurons, with much greater synchrony than normal. The discharge produces the clinical symptoms of seizure, and the loss of the independent neuronal discharge takes away the ability of the brain to function. Hence, with an absence seizure, which is a type of generalized seizure, there is no generalized tonic–clonic motor activity, but the brain cannot solve complex math problems or write books because many neurons are engaged in this synchronous dance.

Focal-Onset Seizure

One part of the brain shows abnormal discharge causing the clinical seizure activity. Because of the mapping of the cortex, discharges in the central part of one hemisphere result in clonic and sometimes tonic activity of the contralateral body or part of it. The part of the brain serving the upper extremity is more commonly affected by pathology producing focal seizures, so the arm and hand are affected by focal seizure activity more often than the leg. However, many focal seizures affect areas other than motor control regions, so they may have a myriad of other symptoms with little or no motor manifestations, including psychic, forced thoughts or other sensation, or disturbance of consciousness without other motor manifestations.

Brain Death

Brain death is cessation of all brain functions. Note that if we record from individual brain cells, we will detect some potentials, but they are not the ones that will produce the complex patterned electrical discharge which is necessary for consciousness.

Electromyography and Nerve Conduction Studies

KARL E. MISULIS AND EVAN M. JOHNSON ∎

HISTORY AND DEVELOPMENT OF ELECTROMYOGRAPHY AND NERVE CONDUCTION STUDIES

Educated people knew since the time of Socrates that muscles could be stimulated to contract by electricity, and this was often thought to have beneficial effects for some disorders, especially when there was paralysis. In the late 1800s, the first recordings were made of electrical activity from contracting muscle. Visualization of the activation was easier after invention of the oscilloscope. Oscilloscopes were available earlier but started to be used for physiological measurements in the 1920s. These were analog devices. Because the traces were dynamic, complicated photographic equipment attached to the scope was required for storage.

The digital transformation of electromyography (EMG) occurred in the 1980s, when conversion not only allowed improved ease of acquisition and storage but also allowed for digital manipulation of the data.

METHODS OF DATA ACQUISITION

Generic aspects of data acquisition are discussed in Chapter 3. Here, we focus on the application of these principles to EMG and nerve conduction studies (NCSs).

Modern EMG equipment consists of the suite of modules within one device. The fundamental elements are as follows:

- Recording hardware and software
- Stimulator
- Computer that drives and coordinates the stimulating and recording function
- Display
- Storage

Recording hardware and software are fundamentally identical to those of electroencephalography (EEG) machines, so a discussion of them is not be repeated here. Briefly, analog amplifiers for the biological signal feed to analog-to-digital converters, which then transfer the digital data to a computer for analysis and display.

Stimulating functions are almost the reverse of the recording methodology. The computer drives a stimulator module, which takes the computer directions to produce an electrical stimulus of the specified polarity, amplitude, duration, and waveform.

EMG as well as most other neurophysiologic equipment has set options that are within certain guardrails. Medically dangerous shocks cannot generally be delivered. Similarly, the options for recording will generally be within appropriate parameters for the study. In fact, the parameter choices are usually specific to the study to be performed. For example, choosing sensory nerve conduction velocity (NCV) will produce different options for stimulation and recording parameters than if we choose motor NCV.

BASICS OF EMG AND NCS PHYSICS

Electromyography and NCSs are the basic components of neuromuscular diagnostics. There are multiple modes of each of these, but the basics are recording from nerves and muscles and stimulating nerves.

Recording electrical activity from muscle is accomplished by electrodes either placed on the skin or inserted into the muscle. Surface electrodes are similar to those described for EEG in Chapter 6, although the composition is different and the method of connection to the skin differs as well. Commonly used electrodes are bar electrodes, in which anode and cathode are separated by a fixed distance and can be placed together on the skin overlying a peripheral nerve or muscle. This provides bipolar recording. When the electrode is over a nerve, the recording is of a compound nerve action potential. When the recording electrode is over a muscle, the recording is of a compound action potential from muscle. This is termed *compound motor action potential* (CMAP), although it is important to recognize that the potentials recorded are from the muscle fibers and not the motor nerves themselves.

Motor Conduction

CMAPs are not recorded from peripheral nerves because there are no purely motor peripheral nerves, although there are purely sensory nerves. If we test conduction between two points on a mixed nerve, we are recording both motor and sensory conduction, and this is not as clinically useful as measuring only motor or sensory.

Because stimulation of a nerve innervating a motor nerve results in activation of the muscle, the time to activation not only includes the conduction time of the nerve but also the time for neuromuscular transmission and time to conduct to the segment of the muscle underlying the recording electrodes. To remove the neuromuscular junction (NMJ) and muscle conduction from the equation, we stimulate two points of the mixed nerve, calculate the difference in time for the CMAP, and divide the distance between the stimulating electrodes by the difference in time of the CMAP to calculate the motor conduction velocity.

Sensory Conduction

To record sensory conduction, we must be cognizant that most nerves are mixed, so we need to use one of three approaches. First, we could test nerve conduction on a nerve that is purely sensory, but there are few of those. Second, we could stimulate the nerve on a digit where there are no motor branches so that we are only stimulating sensory nerves, and then we record proximally on the mixed portion of the nerve. Third, we could switch the electrodes around and stimulate a mixed nerve but record only from the skin innervated by that nerve, and not from near the muscles that the nerve might innervate. This latter method has a higher risk of artifact from muscle activation contaminating the recording, so this is less frequently used.

F-Waves

F-wave is a test of the conduction of motor axons proximal to the stimulation site, as opposed to the CMAP, which is testing motor conductions distal to the stimulation site. The "F" stands for "foot" because initially the F-wave was performed on nerves to the foot, but it is performed on other nerves now. A nerve is stimulated with the electrodes in a direction to optimally activate conduction rostrally, toward the spinal cord. The action potential then depolarizes motoneurons in the anterior horn of the spinal cord extending to the dendrites. The depolarization of the dendrites in turn produces another phase of depolarization of the soma that activates the motoneurons again. This action potential is propagated back to the muscle. The test is called the F-wave rather than the F-reflexes because it is not a reflex; rather, it is a reflection of action potential up and then down an axon and is not the product of electrical stimulation of a reflex pathway.

H-Reflex

The H-reflex is named for Paul Hoffman, a German neurologist (not the same Hoffmann as the Hoffman reflex). The H-reflex tests the response to stimulation of muscle afferents. These afferents are generally large-diameter, low-threshold afferents so that an H-reflex is elicited with lower level stimulation than the F-wave. Classically, the tibial nerve is electrically stimulated behind the knee with the electrodes oriented so the nerve volley is more proximal than distal. As the intensity is gradually increased, the H-reflex appears, but with further increase the H-reflex disappears as the CMAP appears. This is because the low-intensity stimulation produces activation of mainly large muscle spindle afferents, whereas the greater intensity activates alpha motoneurons.

Tests of Neuromuscular Transmission

There are multiple tests of neuromuscular stimulation, including paired stimulation, repetitive stimulation at slow and fast rates, and single-fiber EMG. These all leverage the physiology of the NMJ and the biophysics of muscle and nerve function to reveal deficits in neuromuscular transmission that can be very different depending on whether the patient has myasthenia gravis, Lambert–Eaton myasthenic syndrome, botulism, or a related condition.

Electromyography

Electromyography is direct recording from skeletal muscle using a needle electrode. Usually, the active recording surface of the electrode is not the entire needle electrode surface but, rather, only a small portion of it. The electrode is inserted into skeletal muscle as the patient is at rest. After assessment of electrical activity at rest, the patient is asked to gently contract the studied muscle, and features of the discharge are used to determine whether there is a defect in neuromuscular transmission, loss of motoneurons, or damage directly to the muscle. Additional details of this differentiation are presented later.

NORMAL RESPONSES

Performance of basic NCSs and EMG studies results in predictable findings. This section describes the mechanisms behind the normal response, and the following section describes the mechanisms behind some important abnormal responses.

Compound Motor Action Potential

A normal CMAP is shown in Figure 7.1. This is the summed recording of a volley of action potentials stimulated by electrical current. The electric stimulus produces depolarization of the nerves near the cathode (the negative pole of the stimulator). The anode (positive pole) of the stimulating electrode is a short fixed distance from the cathode, and for motor conduction studies, the anode is proximal to the cathode—that is, the stimulus is generated nearest the cathode and propagates toward the NMJ and muscle from there. For the "compound" portion of the term, multiple motor axons are activated. For each of these, hundreds of muscle fibers are activated, so the recorded potential is the combined response from many muscle fibers.

Sensory Nerve Action Potential

For a normal sensory nerve action potential (SNAP), either the stimulating or recording electrode is over a purely sensory portion of a peripheral

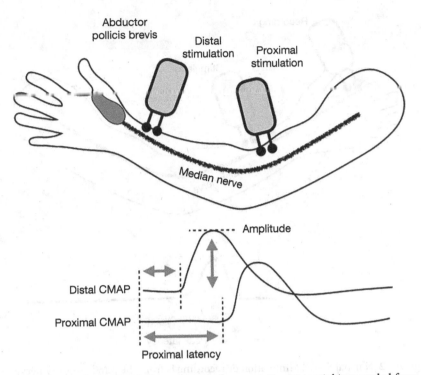

Figure 7.1 Normal CMAP. Median compound nerve action potential is recorded from the abductor pollicis brevis. Stimulation is in two locations, one in the lower part of the upper arm and the second in the wrist. Conduction velocity is calculated as the difference in time to activation from the two stimulation sites divided by the distance between the sites.

nerve, and it is the summed potentials of hundreds of sensory axons. A diagram of electrode positions and sample recording is shown in Figure 7.2. This is a nerve recording, so it is generally faster and smaller in amplitude than the motor action potentials from muscle.

Motor Nerve Conduction Velocity

Motor nerve conduction velocity is the speed of conduction of the fastest of the motor nerves. Because the largest amount of time from nerve activation to muscle fiber electrical activation is not the motor nerve, we use two stimulation points and calculate the difference in time to muscle

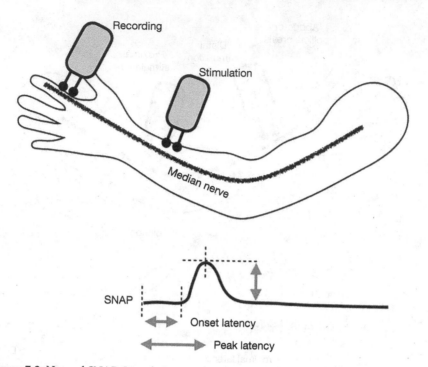

Figure 7.2 Normal SNAP. Stimulation or recording is from the purely sensory nerves of the finger because there are no median innervated muscles in the finger. Note that because this does not depend on muscle activation, the stimulating and recording electrodes can be reversed.

activation, dividing that into the distance between the stimulating electrodes.

Sensory Nerve Conduction Velocity

Sensory nerve conduction velocity is the speed of the fastest nerves, so the measurement is to the onset of the SNAP.

F-Wave

F-wave latency is variable from trial to trial because of differences in activation of the intraspinal portion of the neuronal activation. Therefore, it

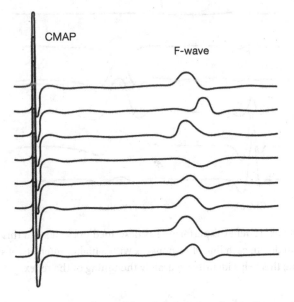

Figure 7.3 F-wave. F-wave is performed by stimulation and recording as described in the text with a series of waves to determine the response.

is measured as the shortest F-wave latency in 10 trials. A sample recording is shown in Figure 7.3.

H-Reflex

H-reflex is elicited by gradually increasing stimulus intensity and watching for the response in the appropriate time window. As the stimulus is increased, the response appears, grows, declines, and disappears. Measurement is of the shortest latency in the train of stimuli. A sample recording is shown in Figure 7.4.

EMG

Electromyography of multiple muscles is usually performed, and there are some basic features of the normal EMG. Note that some of the terms used have audio implications because an important part of analysis is audio; we

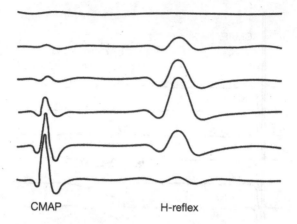

CMAP   H-reflex

Figure 7.4 H-reflex. H-reflex is performed as described in the text. In this series of stimuli, from top down, each line of response is with a higher intensity of stimulus than the previous line; this helps identify accurately the timing of the reflex.

listen to the recorded electrical activity through a speaker as well as look at the potentials on the computer screen. Figure 7.5 shows some normal tracings from an EMG epoch. The findings in a normal muscle are as follows:

- *Rest:* Muscle is quiet, with few muscle discharges. There is some low-amplitude rumble of the baseline.
- *Mild activation:* Multiple motor unit potentials (MUPs) are seen.
- *Maximal activation:* Innumerable MUPs overlap so that one cannot distinguish between them.

ABNORMAL RESPONSES

The spectrum of abnormal responses in NCV and EMG is extensive, and the purpose of this discussion is to explain the physics and related physiology of the findings, so we focus on some of the most common and more important abnormalities. Rather than organize the findings by specific test, we discuss how particular types of disease produce abnormalities. Figure 7.6 illustrates a few of the abnormalities encountered, with a focus on neuropathy and myopathy.

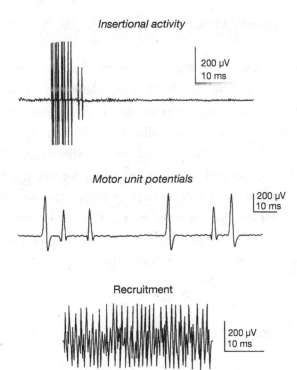

Insertional activity

200 μV
10 ms

Motor unit potentials

200 μV
10 ms

Recruitment

200 μV
10 ms

Figure 7.5 EMG. Insertional activity occurs as the needle is advanced in the muscle. Motor unit potentials are where the patient makes a mild contraction so that a few units can be examined. Recruitment is obtained when the patient makes a maximal contraction so it can be determined if activation is of normal pattern.

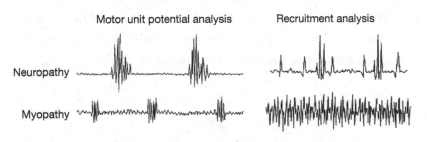

Motor unit potential analysis   Recruitment analysis

Neuropathy

Myopathy

Figure 7.6 Abnormal EMG findings. With neuropathy, the motor units are high-amplitude polyphasics because of reinnervation of denervated muscle fibers, and with recruitment there are fewer but larger units. With myopathy, the motor units are smaller because of failure of electrical activity of many muscle fibers, and recruitment pulls in more motor units because each is weaker than normal.

Active Denervation of Muscle

Muscle is quiet at rest, but with active denervation there is instability of the muscle membrane resting membrane potential so that it is depolarized, and oscillations in this potential occasionally reach threshold. Therefore, there is activation of single muscle fibers by this seemingly random fluctuation. This produces fibrillations and positive sharp waves that are from the same pathology but with different recording geometry. With active denervation, there is also a reduction in the number of motor units, so with maximal contraction, the mass discharge, termed interference pattern, is reduced. However, when the discharges of individual motor units are reviewed with mild muscle contraction, they appear fairly normal—few phases with short overall duration.

Chronic Denervation of Muscle

With chronic denervation, the membrane instability of active denervation has largely disappeared, so fibrillation potentials are not prominent. A reduction in the number of active motor units will result in reduced numbers of active motor units. In addition, the motor units will have a different appearance because there has been an attempt at reinnervation of denervated muscle fibers and those reconnections are not as good as those that developed early in life. Therefore, the motor unit potentials are longer and characterized by multiple phases due to the difference in timing from these new connections. These are polyphasic potentials.

Axonal Neuropathy

Axonal neuropathy usually affects both sensory and motor nerves so that there is a reduction in the amplitude of the SNAP but little change in sensory NCV. Motor NCV shows reduction in amplitude because of denervation, resulting in lesser activation of muscle fibers. There is also some dispersion of the action potentials because of attempts at reinnervation.

EMG shows active denervation if acute, chronic denervation if not acute and no new denervation is occurring, and both if it is a progressive neuropathy with active and chronic neuronal damage.

Demyelinating Neuropathy

Demyelinating neuropathy produces dispersion in both motor and sensory action potentials, and a slowing of conduction, because of the damage to the myelin sheath. Because of this dispersion, the amplitude of the potentials is also lower, but the area under the curve should be approximately normal.

Motor Neuron Degeneration

Motor neuron degeneration is a purely motor event and the classic disorder is amyotrophic lateral sclerosis, in which there is degeneration of the motoneurons as well as degeneration of the corticospinal tracts driving them. Here, we focus on the lower motoneurons. Denervation of muscle fibers results in fluctuations in membrane potential that occasionally reach threshold, producing muscle fiber action potentials. These are single muscle fibers, so they do not have the amplitude and configuration of motor unit potentials. Fibrillations and positive sharp waves are two manifestations of the same muscle fiber activity but are due to different recording geometries.

Neuromuscular Transmission Defects

Conditions that affect NMJ function produce complex findings which depend on the type of defect. However, these are generally distinguished by insecurity of NMJ transmission. This results in changes in response from the muscle when multiple motor nerve stimuli are delivered. This can be repetitive stimulation at various frequencies or paired stimuli.

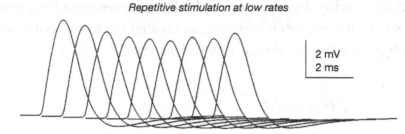

Figure 7.7 Neuromuscular junction defect. Myasthenia gravis can result in a decremental response to repetitive stimulation. This is because there is failure of transmission at some of the neuromuscular junctions.

Depending on the type of NMJ defect, rapid delivery of motor nerve stimuli can result in augmentation or degradation of the muscle fiber electrical response. Figure 7.7 shows a decremental response from a patient with myasthenia gravis.

Myopathy

Myopathies produce instability of the muscle fiber membrane, giving some of the same abnormal fibrillation potentials that are seen with active denervation. Motor units are quite different, in that there are small, low-voltage polyphasics with a short duration rather than the large-amplitude polyphasics of denervation. This is because of reduction in the number of muscle fibers with each motor unit without dispersion due to reinnervation. With maximal effort, myopathic muscles produce a large amount of low-voltage activity because there are fewer muscle fibers per motor unit but not fewer motor units, as there are in neuropathic dysfunction.

CLINICAL CORRELATION

Normal

- Motor conductions: normal CMAP amplitude and NCV

- Sensory conductions: normal SNAP amplitude and NCV
- EMG: no denervation, normal MUPs, normal recruitment

Conductions of motor and sensory axons are normal, and there is no denervation to cause EMG changes. Normal study indicates that the nerves from the spinal cord to the muscles and skin are intact. It does not report the status of the higher motor and sensory centers.

Guillain–Barré Syndrome

- Motor conductions: CMAP dispersion, NCV slowed, F-wave delayed
- Sensory conductions: SNAP delayed, NCV slowed
- EMG: usually no denervation unless severe and even then not acutely

Early changes pathologically are demyelination resulting in slowed motor and sensory nerve conductions. Demyelination produces loss of the normal high-velocity saltatory conduction, jumping from node to node. This not only slows conduction but also increases the chance of failure of transmission. Hence, amplitude is reduced in addition to velocity. When there is failure of transmission, blocking occurs because the depolarization at the dysfunctional node is insufficient to generate an action potential that can be propagated further. Reduced synchrony because of uneven demyelination between axons results in dispersion of the waveforms. EMG usually shows no denervation because at least initially, the motor nerves are intact, even if they do not conduct properly. Eventually, with motor axon dysfunction of sufficient severity and duration, denervation can occur, and active denervation can be seen.

Carpal Tunnel Syndrome

- Motor conductions: median distal latency prolonged through the wrist; otherwise nerves normal
- Sensory conductions: median sensory conductions slowed through the wrist
- EMG: usually normal, but may show denervation in abductor pollicis brevis

Entrapment of the median nerve at the wrist produces slowing of especially sensory and often motor conduction through the wrist. Pressure on the median nerve results in depolarization of other nerves. This effect does not affect all nerves equally, so there is often reduction in amplitude because of dispersal of the waveform: The same number of axons conduct, but the waveform is spread out due to lesser synchrony. With more severe entrapment, the compression is sufficient to depolarize the nerves enough to produce conduction block, thereby further reducing amplitude. Proximal median motor conductions are normal, and conduction of other nerves is normal in the absence of other pathology. EMG is often normal; however, in more severe cases, denervation is seen, but only in distal median-innervated muscles. Median-innervated muscles in the forearm are unaffected because they are unaffected by the compression; there is no neighbor effect of damage.

Myasthenia Gravis

- Motor conductions: normal routine conduction
- Sensory conductions: normal routine conductions
- EMG: routine EMG usually normal; single-fiber EMG shows increased jitter
- Repetitive stimulation: decremental response at low rate of stimulation

Myasthenia gravis is caused by antibodies to the acetylcholine receptor, which results in internalization and degradation of the receptor on the muscle fiber. As a result, a single action potential may activate most muscle fibers, but repetitive stimulation results in failure of activation for some fibers—hence the decremental response. Single-fiber EMG is able to show details of activation of individual muscle fibers. Although normally they should be almost perfectly time-locked, there is more synaptic insecurity with myasthenia and therefore more variation in timing of activation of individual muscle fibers and often failure of certain muscle fibers to be activated.

Muscular Dystrophy

- Motor conductions: normal
- Sensory conductions: normal
- EMG: myopathic changes of affected muscles

Muscular dystrophy is degeneration of the muscle fibers, typically due to metabolic changes within the muscles rather than due to damage to the innervating axons. Because the neuronal components are not the primary targets of the pathology, motor and sensory conductions are normal, at least with milder or earlier disease. With more severe muscle degeneration, the muscle fibers may be unable to sustain the action potentials required for motor conduction studies. Sensory conduction will be normal unless there is sensory nerve compression from patients with severe paralysis. EMG shows myopathic changes—that is, fibrillation potentials due to spontaneous discharge of affected muscle fibers. Positive sharp waves are potentials of the same pathophysiologic origin but with a different recording geometry. MUPs are smaller because of failure of some muscle fibers to fire, and they are polyphasic because of reduced synchrony as well as dropout of muscle fibers. With increasing muscle effort, maximal recruitment activates more muscle fibers than normal. Because each muscle fiber has less power, more have to be recruited.

Evoked Potentials

KARL E. MISULIS ∎

HISTORY AND DEVELOPMENT OF EVOKED POTENTIALS

Evoked potentials (EPs) are the electrical responses to stimulation. Most EP studies are sensory, with visual, auditory, and somatosensory being the principal modalities. There are also motor evoked potentials.

The earliest EP recording that we are aware of was in 1875 when Richard Caton at the Royal Infirmary in Liverpool recorded cerebral potentials in response to sensory stimulation in the periphery. The following year, he found that potentials could be identified in response to light stimulation of the eye. This was reproduced and advanced elsewhere. Although EPs would not be used clinically until much later, it became common knowledge that activation of various modalities of sensory afferents produced electrical responses from the brain and that even early technology could record these potentials.

However, these potentials were irregular and inconsistent, which made them of limited use clinically because only integrity of a system could be determined, not degrees of involvement. Clinical utility of EPs exploded when digital conversion of these responses allowed for averaging of many trials. This resulted in the ability to generate standard consistent reproducible responses and normative values that could be used for comparison study.

EPs gained increasing use as diagnosis of central disorders was partnered with more therapeutic options. Multiple sclerosis (MS) was no longer solely a clinical diagnosis; rather, there was objective evidence of demyelination that could be documented and measured. EP could show lesions that computed tomography could not see.

EPs were used in suspected cases of MS not only to reveal objective evidence to correlate with subjective clinical findings but also to identify clinically silent lesions.

The use in patients with hearing loss was to help differentiate between localization of lesions, and that information could have profound therapeutic implications. If the auditory EP suggested an acoustic nerve lesion, then surgery might be an option. If a brainstem lesion was suspected from the findings, then direct intervention was less likely to be initiated without worsening deficit.

The advent of magnetic resonance imaging (MRI) has made the utility of EP for diagnosis of especially demyelinating disease less important. MRI is better able to show the extent of demyelinating and other structural diseases even when there is no measurable objective clinical evidence of dysfunction.

Currently, sensory EPs are used mainly for intraoperative monitoring, although they are still occasionally used when additional functional information is needed and for patients who cannot undergo MRI. In our current society, there seems to be an increasing proportion of patients who harbor ballistic fragments, making MRI riskier and sometimes impossible.

Motor EPs are seldom used in neurologic practice. This is where there is electrical or magnetic stimulation of the brain, causing activation of descending motor pathways and muscle. Historically, this was first done electrically by Fritsch and Hitzig in 1870, showing that electrical stimulation of the cortex in dogs produced motor activity. Penfield and his team took this to another level with mapping by intraoperative electrical stimulation. Currently, motor activity is usually generated by transcranial magnetic stimulation. Although this is sometimes used for intraoperative monitoring, it is used more often as a research tool.

PHYSICS, PHYSIOLOGY, AND PERFORMANCE

Evoked potentials are so called because they use a stimulus and a response. The stimulus can be any sensory input. Somatosensory EPs have the closest relationship to electromyography and are elicited by electrical stimulation of a peripheral nerve and recording from nervous structures extending from proximal nerves to spinal cord, brainstem, and then cerebrum cortex. The upper and lower extremities can be studied, and by using response analysis, it can be determined whether there is slowing of conduction through any segment of the pathway.

We consider the following types of EPs:

- Visual EP
- Somatosensory EP
- Auditory EP
- Motor EP

Visual EP is produced by recording from the scalp while delivering a visual stimulus that is either flashes of light or a patterned stimulus delivered to one or both eyes and to one or two hemifields.

Somatosensory EP is produced by stimulation of a peripheral nerve that contains sensory afferents and recording from usually multiple locations. Among the regions that can be recorded are a more proximal portion of the nerve to ensure integrity of the stimulus, the brachial or lumbosacral plexus depending on location of the stimulated nerve, spinal cord segment(s), and brain.

Auditory EP is produced by recording from the scalp while delivering sounds to one or both ears. The recording assesses conduction in the nerve and brainstem especially.

Motor EP is produced by stimulation of the cortex and recording from descending axons and muscles. Although electrical stimulation can be done, there is concern that repeated electrical stimulation of the cortex might have adverse effects. Therefore, magnetic stimulation is most commonly used. Years of study have shown that the risk of this stimulation is low. For this chapter, we do not consider motor EP further because we do not routinely use this modality in neurology clinical practice and have little experience with it.

Interpretation of EPs relies on having accurate normative data, and that data need to be appropriate not only to the nerve(s) being studied but also to the dimensions of the patient. Therefore, a fairly rigorous standard of methodological implementation is used.

Performance of EPs generically involves stimulation in the particular modality and recording from the various regions of the body surface appropriate to the stimulus. The responses are low amplitude and often variable from trial to trial. Therefore, averaging is essential for most of these. A single trial may give a waveform that could be interpreted, but because there is much inter-trial variability, hundreds and sometimes thousands of trials are needed. Because of the low amplitude of the responses and the multiple trials needed, often some trials are contaminated by high-amplitude artifact. A single trial of this magnitude could distort an averaged waveform, so artifact rejection is used, as described in Chapter 2. Artifact rejection occurs automatically, where the acquisition software is programmed to reject trials that are judged artifact.

Neural Generators of Evoked Potentials

When we give a stimulus and record a response from a peripheral nerve, it is easy to identify the generator—it is the volley of action potentials in the nerve. Similarly, when we stimulate a motor nerve and record the motor action potential, it is clear that the response is the summed action potentials of the muscle. However, with more proximal recordings, it is not clear exactly where the potentials are coming from. Part of this is that we make the transition from near-field potentials to far-field potentials. Part is that we can get anchored on the function of specific regions, especially nuclei. We might assume that a particular wave is from that nucleus or another, but in reality, nuclei usually do not produce the summed electrical activity in a vector that is recordable from the surface of the body. Unless an electrode is in the nucleus, we cannot record its electrical activity. Even if there is a linear array of afferent and efferent axons into and out of a nucleus, central synapses are not arranged along that axis. They are organized around the dendrites and soma of the next-order neurons so that although there is significant charge movement with synaptic transmission, we are not able to see it from distant volume-conducted recordings.

What we can record is electrical activity in a volley of axons that are generally oriented in one direction. This means that ascending volleys of impulses in hundreds of axons in the spinal cord or thalamocortical radiations is much more likely to be what we record than what the spinal nuclei or thalamus actually generates.

Auditory Evoked Potentials

Auditory evoked potentials (AEPs) were historically called brainstem auditory evoked potentials, but the "brainstem" portion was dropped because the response is not just from the brainstem. This is tested by delivering stimuli given through headphones to one or both ears and then recording from the scalp. When one ear is tested, there is commonly a masking sound given to the opposite ear to cover the possibility that the unintended ear might perceive some of the stimulus delivered to the tested ear. Regardless of reason, both ears are tested for comparison.

There is a complex waveform that has been at least partially correlated with neuroanatomical structures. Refer to Figure 8.1 during this discussion.

The most important waves used for interpretation are I, III, and V. Wave I is generated by the distal acoustic nerve and is believed to be the afferent volley. Identification of this is important to ensure that there was stimulation and transduction and also to use this latency as a reference to calculate conduction times to structures generating later waves. From Figure 8.1, it is apparent that there is wave I only from the stimulated side.

Wave II is thought to be generated by the proximal (closer to the brainstem than wave I generation) portion of the acoustic nerve. This is not

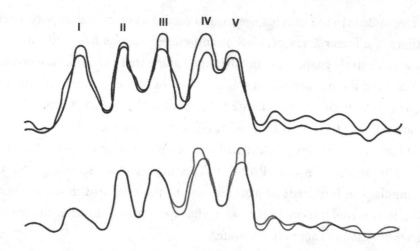

Figure 8.1 Auditory evoked potential. Stimulated ear is on top and nonstimulated ear is on the bottom. Waves I–III assess conduction from acoustic nerve through lower brainstem, and waves III–V assess conduction through upper brainstem.

certain. Because of this uncertainty and because wave I is more reliably recorded, wave II is not used for interpretation.

Wave III is thought to be generated by projections from the superior olive through the lateral lemniscus. This is used for interpretation and is thought to represent activity in the lower brainstem.

Wave IV is often fused as part of a IV–V complex, and it might be from the ipsilateral lateral lemniscus, which relays information from the cochlear nucleus to the contralateral inferior colliculus in addition to other brainstem nuclei. Wave IV is not used for interpretation.

Wave V is likely generated by projections from the pons to midbrain and as such is likely a composite with wave IV of these projections. Wave V is most consistent and used for clinical interpretation.

There are other waves later than those discussed, but they are not used for clinical interpretation.

By assessing the response to auditory stimuli, we can determine whether there is damage between the ear and the brainstem, between the lower brainstem and the upper brainstem, or whether there is a combination of lesions. This has been used in the past for MS evaluation, but it is no longer used. Auditory EPs are now used predominately for children to assess hearing prior to our ability to ask the patient.

The following measurements are made for interpretation:

- Wave I latency
- Wave III latency
- Wave V latency

On the basis of these data, we calculate the following:

- I–III interpeak interval
- III–V interpeak interval
- I–V interpeak interval

Interpretation of findings based on the localization of generators and pathways is as follows:

- Increased wave I latency: dysfunction of the distal acoustic nerve
- Increased I–III interpeak interval: dysfunction of the pathway from the acoustic nerve to the inferior pons
- Increased III–V interpeak interval: dysfunction in conduction between caudal pons and midbrain

There are more complexities in interpretation with some more uncommon presentations, but the point is that the AEPs can help localize lesions of the acoustic nerve and brainstem.

Visual Evoked Potentials

Visual evoked potentials (VEPs) are delivered by flash or checkerboard stimuli shown on a screen in front of one or both eyes, with the left or the right or both hemifields stimulated. The purpose of VEPs is to identify lesions in the visual pathway that can be of the optic nerve prior to the

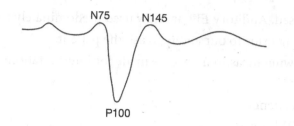

Figure 8.2 Visual evoked potential. The P100 is used for routine analysis. Note that the figure shows negative up and positive down. Many displays have this reversed so that the wave of interest is upward, more evident to humans.

optic chiasm, behind the chiasm extending into the lateral geniculate, the optic radiations, or the visual cortex. Responses to the visual stimuli can help determine the localization of these lesions. Figure 8.2 shows a sample of a normal VEP.

The stimuli differ depending on the condition of the patient and the purpose of the study. One of the most common uses of VEP is assessment of visual pathways in children. For this indication, flash stimuli are delivered to each eye and recordings are made from the posterior scalp, over the occipital visual cortex. The flash stimulus does not depend on patient cooperation and can even be done on sedated patients.

For adults, pattern-reversal stimulation is usually done. This is a checkerboard pattern with an adjustable check size. The bright and dark checks are reversed at a specific rate. One of the benefits of this pattern of stimulation is that there is no overall change in illumination; it is purely a change in which retinal cells are activated. However, in order for this to be performed, there has to be cooperation from the patient to fixate on the target. One hemifield can be tested at a time as well, so between the different modalities, dysfunction in pre- and post-chiasmatic function can be evaluated.

Measurements are the latency of a potential recorded over the occipital cortex, with positivity at approximately 100 msec latency, so this is termed P100. This is the wave used for interpretation. There are negative waves prior to and after this, which are labeled for their estimated normal latency: N75 and N145.

Abnormalities of the pattern-reversal VEP are interpreted as follows:

- Prolonged P100 with monocular stimulation—defective conduction in one optic nerve on the same side as stimulation
- Prolonged P100 with binocular hemifield stimulation—defective conduction behind the optic chiasm on the opposite side to stimulation

Flash VEP is less sensitive for determining the integrity of the pathways and is used now mainly for evaluation of preterm infants, where abnormalities have some prognostic value for later impairment in neurologic function and even survival.

Somatosensory Evoked Potentials

Somatosensory evoked potentials (SEPs) are produced by stimulation of peripheral nerves and recording from more proximal portions of those nerves, over the spinal cord, and over the brain. SEPs can be performed with stimulation of the arm or leg, and stimulation is of a major nerve. For example, SEP from median nerve stimulation would place the stimulating electrode over the median nerve at the wrist, and recording electrodes would be at several locations. During this discussion, refer to Figure 8.3, which shows a normal median SEP. Recording at Erb's point above the clavicle would assess conduction from the wrist to the brachial plexus. Recording electrode over the cervical spinous process assesses conduction through the proximal plexus to the spinal cord. Recording electrode from the contralateral cerebral cortex allows evaluation of conduction through the brainstem, thalamus, and the cortex. Note that the potentials are not of individual neurons but, rather, the vector sums of waves of activation of the projections. It is intellectually attractive to think of each wave as being created by specific nuclei, but the neurophysiology is not so accommodating.

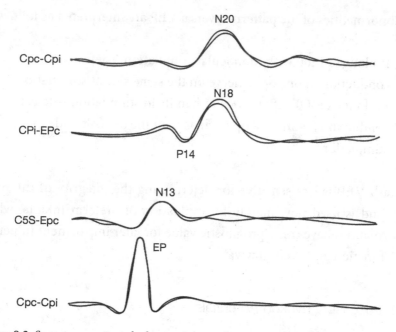

Figure 8.3 Somatosensory evoked potential—median nerve. The median nerve is stimulated, and recordings are made from the brachial plexus starting at the lowest trace. The higher traces are from neuroanatomically higher tracts, as detailed in the text.

Figure 8.3 shows normal recording of median nerve SEP. These are data from one side. As displayed, the channels are as follows:

- CPc–CPi: centroparietal contralateral–centroparietal ipsilateral
- CPi–EPc: centroparietal ipsilateral–Erb's point contralateral
- C5S–EPc: cervical spine 5–Erb's point contralateral
- EPi–EPc: Erb's point ipsilateral–Erb's point contralateral

The waves used for interpretation are identified by a letter indicating location or polarity and a number representing expected latency. After stimulation of the median nerve, the major waves are as follows:

- N9: identified at Erb's point (EP) approximately 2 or 3 cm above the clavicle and representing conduction through the upper trunk of the brachial plexus

- N13: ascending conduction through the cervical spine
- P14: also ascending conduction through the cervical spine, and which was classically used for interpretation
- N20: scalp potential from thalamocortical radiations

From these values, intervals are calculated that represent times for conductions. From these, the following interpretive assessments of abnormalities can be made:

- Delayed N9 with normal N9–P14 and P14–P20: lesion of the somatosensory nerves at or distal to the brachial plexus
- Increased N9–P14 interval with normal P14–N20 interval: lesion between Erb's point and the lower medulla
- Increased P14–N20 interval with normal N9–P14 interval: lesion between the lower medulla and the cerebral cortex

Similar techniques and interpretive analyses can be made for lower extremity SEPs, but this discussion is intended to illustrate the physics and physiology of the technique rather than teach the study itself.

DATA ACQUISITION AND ANALYSIS

Because EPs are very small, they have to be averaged so they can rise above electrical and physiological noise. Therefore, hundreds and sometimes thousands of trials are used in order to get a response.

First, we need a very high-gain amplifier system, beginning with an analog component attached to the electrodes on the patient. The analog signal is then fed to the analog-to-digital converter, which then feeds the signal to the computer for analysis. The trials are averaged, and usually a running average is visible on the screen.

Because of the low amplitude of the responses, minor movements and electrical transients can create significant artifact in the traces for individual trials, so the EP systems have methods for removing trials from the average contaminated by this artifact.

Analysis is a combination of automatic and manual. Most modern acquisition devices make good estimates of the locations where the measurement markers should be. This is similar to the automatic marking of nerve conduction studies. However, it is not uncommon for the markers to be adjusted by the technician and sometimes by the reading clinician because there are significant irregularities in the traces, as is evident from the samples shown. Small adjustments in marking can result in significant changes in interpretation.

CLINICAL UTILITY OF EVOKED POTENTIALS

Evoked potentials were historically used to evaluate physiological processes of the brain and spine when MRI did not provide the details on structure and health of cerebral and spinal structures that it does today. Subtle lesions from MS could be missed on early imaging but could produce findings on EPs. Although this function has waned, EPs are still used, but less so.

AEPs are used mainly for assessment of hearing in newborns when we cannot clinically assess hearing, for intraoperative monitoring during acoustic neuroma surgery, for evaluation of patients considered for cochlear implants, and for assessment of various patterns of hearing loss.

VEPs are used to assess the pathways from the eyes to the brain, mainly to assess whether the signals from the eyes are being conducted to the occipital cortex. VEPs can help identify subtle lesions with optic neuritis because this is not always seen on MRI with damage from demyelinating disease. Also, VEPs can assess conduction in patients with optic nerve gliomas or neurofibromatosis. Intraoperative VEP is used to assess and minimize visual deficit in surgeries when the optic nerves are at risk of damage.

SEPs are used mainly for intraoperative monitoring especially during spine surgery to ensure that the manipulations are not adversely affecting physiologic function at a time when the patient cannot be examined clinically. Also, SEPs are sometimes used for prognostication after cardiac arrest when brain damage is not complete, resulting in brain death, but there is such damage that prognosis is poor for meaningful survival.

CLINICAL CORRELATION

Thoracic Spinal Cord Injury

Damage from almost any cause, such as transverse myelitis, tumor, or trauma, can result in abnormal SEP with lower extremity nerve stimulation (tibial and peroneal) yet normal function with upper extremity SEP (median). If the SEP is performed during spinal surgery, distortion, delay, or loss of the wave can indicate risk of intraoperative cord injury.

Optic Neuritis

Damage to the optic nerve behind the eye and before the optic chiasm produces increased latency of the EP. In a patient who had known optic neuritis, this is expected, but if an abnormal response is seen in a patient without known involvement of that eye, then a subclinical event is suspected.

Acoustic Neuroma

There are multiple waves comprising the AEP. Acoustic neuroma is a tumor affecting CN8 before reaching the brainstem. AEP shows prolongation of the waves representing this interval, specifically waves I and III. If the damage to the nerve is severe, then there may be no response from the brainstem, so there are no waves after wave I.

Ultrasound

EVAN M. JOHNSON ■

WHAT IS ULTRASOUND?

In simplest terms, *ultrasound* is sound waves at frequencies higher than a human can hear. Infrasound, by contrast, is sound frequencies below human perception. The typical human hearing frequency range is approximately 20 Hz to 20 kHz, although the high end diminishes after childhood toward 15 kHz. Keep this in mind when shopping for your next pair of speakers or headphones!

Sound waves are not part of the electromagnetic spectrum. This spectrum, from radio waves and microwaves to gamma rays, is composed of energy particles traveling at the speed of light (299.8 million meters per second), whereas sound is far slower (343 meters per second in dry air).

The speed of sound is variable and dependent on the medium in which it is traveling. Sound cannot exist in a vacuum, no matter how cool science fiction movies sound as spaceships blast one another. Sound is not composed of an independent particle/photon but, rather, is a wave formed by the oscillating motion of particles already making up a medium. The properties of the medium determine the speed at which sound travels: The speed of sound in dry air, as mentioned, is approximately 300 m/sec; in seawater it is approximately 1500 m/sec; and through steel it is approximately 5900 m/sec. For a given temperature, the speed of sound through a specific medium is calculated by taking the square root of the coefficient of stiffness divided by density. Sound conventionally travels quickest through a stiff medium with lower density.

As per the song titled "The Boxer" by Simon & Garfunkel, "a man hears what he wants to hear, and disregards the rest," and so it was with ultrasound in history. Scientists justifiably failed to appreciate light and sound outside the range of human perception for several millennia, making them relatively late discoveries in the annals of scientific history. The first steps toward the discovery of ultrasound are generally considered to have come per Lazzaro Spallanzani in the late 1700s. Spallanzani, born in Italy, had a fascinating life story, becoming both a Catholic Priest and a devout biologist with numerous discoveries funded by the Church.

Among them, Spallanzani took notice that bats demonstrated a fairly unique ability to fly in darkness. In multiple experiments, several of which would be deemed cruel today, he showed that bats could navigate even obstacle-laden courses without the benefit of light or eyesight but that a deafened bat could not. Francis Galton would help establish the existence of sound outside the range of human detection with his animal studies that led to the creation of the dog whistle in 1876. Later work by Donald Griffen in New York during the World War II era showed that bats could not travel if their mouths were closed shut and that captured sound generated from bats was above frequencies of human detection; he ultimately coined the phrase "echolocation" in 1944.

HOW DID SONOGRAPHY COME ABOUT?

The first steps toward sonography, or imaging via sound waves, required the controlled generation of sound. The study of sound took place over the majority of human history, but it was the 19th-century works in electricity that propelled matters forward. Electricity had been observed for millennia, but the discovery of harnessing and generating electrical currents revolutionized science and technology. The realization of interplay between magnetism, light, and electricity paved the way for conversion of energy forms: radio, speakers, light bulbs, etc.

Piezoelectricity followed in these footsteps. *Piezo,* meaning press, electricity is the interconversion of electricity and mechanical deformation. The concept was first explored in the 1700s by Carl Linnaeus, Franz Aepinus René, Just Haüy, and Antoine César Becquerel using temperature and mechanical stress. The brothers Pierre Curie and Jacques Curie of radioactivity fame would eventually establish in 1880 that certain solid crystals, most notably quartz and Rochelle salt, could generate electricity through their deformation. A year later, Gabriel Lippmann showed that the reverse was possible.

In subsequent years, this discovery was incorporated into the invention of sonar. The sinking of the Titanic due to the lack of an early warning system for obstacles at night and the advent of the war-ready submarine prompted the need for an underwater detection technique. Radio waves had been shown to be useful for this principle above ground and eventually led to radar (*r*adio *d*etection *a*nd *r*anging), so named in 1940. The British Navy's Anti-submarine Division would produce an early, quartz-based piezoelectric system to both produce and then detect reflected sound waves underwater in 1917. This paved the way for sonar (*s*ound *na*vigation and *r*anging). Today, numerous practical applications make use of piezoelectricity—for example, the electric–acoustic guitar, the cigarette lighter in vehicles, quartz watches, and many more.

The principles behind sonar technology fueled interest in medical imaging applications. In the 1940s, using more advanced transducer–receiver systems, efforts were made to establish the use of ultrasound techniques to non-invasively image human tissue and pathology, which would eventually become an indispensable tool for modern physicians.

Medical sonography devices were first developed in the same concept as X-ray systems, in which a detector sat opposite a ultrasound generator to capture sound waves that passed through an object and "shadows" marked areas of high attenuation. In the 1940s, this concept shifted toward the capture of reflected waves, and a detector and transmitter were located on the same side of the patient. In the mid-1940s, Floyd Firestone produced a system in which a single transducer alternatively acted as both the transmitter and the detector by using pulsed waves.

George Ludwig developed A-mode and demonstrated in the late 1940s that ultrasound imaging could detect gallstones. A-mode, or amplitude modulation, is a single-transducer, single-line form of ultrasound plotting the boundaries seen with corresponding depth.

The echocardiogram was established in 1953 by Inge Edler along with Hellmuth Hertz. Ian Donald pioneered the use of sonography in obstetric and gynecologic patients starting in the mid-1950s, and its use greatly expanded following the advent of B-mode in the 1960s. B-mode, or brightness modulation, is the more familiar two-dimensional ultrasound mapping made possible by multiple transducer elements sending numerous parallel sound waves for echoes.

Baker, Watkins, and Reid helped design pulsed Doppler ultrasound in 1966 to pave the way for functional vascular applications. Doppler employs high-frequency sound waves in order to measure moving particles, such as red blood cells.

WHAT SETS SONOGRAPHY APART FROM OTHER IMAGING MODALITIES?

Sonography is an integral component of medical imaging in virtually all health care facilities today, alongside X-ray imaging, computed tomography (CT), and magnetic resonance imaging (MRI). These imaging modalities offer unique advantages and limitations that promote incorporation of all together for complementary purposes.

Ultrasound offers several specific advantages. It is relatively low cost, both in initial capital investment and in upkeep, making it accessible throughout the world. Whereas a new ultrasound system carries a purchase value equivalent to a new automobile, new CT and MRI scanners have a price tag equivalent to an expensive house in a gated community. Ultrasound is also portable, not requiring a specifically built facility room. Sonography is flexible, allowing an operator to freely image any area of the body of interest. Systems today tend to offer numerous transducer probes to allow optimization of specific exams. Importantly, sonography does not employ ionizing radiation, reducing risk of tissue damage, nor does it employ a magnetic field, removing potential limitations in patient selection. Whereas CT and MRI are sensitive to motion artifacts, ultrasound is not only permissive but also can capture moving structures for important data, such as in echocardiography or vessel Doppler.

Factors that do need to be considered with ultrasound include limitations in depth, magnified with patients of large habitus, and artifacts. Signal-to-noise ratio is not as ideal as in other imaging modalities, and spatial resolution is dependent on several factors, including depth of field. There is a potential for ultrasound waves to heat tissue if applied inappropriately to one area for a long duration. Whereas CT excels at differentiating between tissue types of significant radiopacity, particularly bone and lung from soft tissue, these areas of high reflection or absorption limit ultrasound use.

HOW DOES SONOGRAPHY WORK?

The Physics of Sound Waves

As we delve into the physics of sound transmission, absorption, and reflection, the impedance of the varying materials is paramount. This is a good moment to review the physics of sound.

Sound is an oscillating pressure-based wave that requires a medium with at least some elasticity in order to travel. Sound cannot travel in a vacuum; it moves slowest in gas and fastest in solids. The speed of sound through air is approximately 343 m/sec, it is almost 1,500 m/sec through water, and it is roughly 12,000 m/sec through a diamond. Through gases and liquids, sound travels as a compression wave, where the oscillations are antero- or retrograde to the wave. Through solids, sound can travel as a shear wave, in which the oscillation is perpendicular to the wave. This is visualized by a vibrating string of a musical instrument or a tuning fork.

The speed of sound through a specific fluid is given by the Newton–Laplace equation:

$$c = \sqrt{\frac{K_s}{\rho}}$$

where c is speed of sound, K is the coefficient of stiffness or the bulk modulus, and ρ is the density.

Increased stiffness and/or decreased density results in a higher speed of the wave through the fluid, and vice versa.

In other sound waves, the concept of acoustic dispersion needs to be considered. In certain media, a complex sound wave will be separated into the component frequencies that make it up. In sonography, a wave of a singular frequency is employed, so this effect is not a factor.

The speed of sound through a homogeneous solid is given by the following equation:

$$c = \sqrt{\frac{K_s + G}{\rho}}$$

where c is the speed of sound, K_s is the coefficient of stiffness or bulk modulus, G is the shear modulus, and ρ is the density.

The key difference in the equation for solids is the need to account for the effects seen in a shear wave, which is unique to solids as opposed to fluids or gases.

The speed of sound through a stiff solid such as a metal rod is given by the following equation:

$$c = \sqrt{\frac{E}{\rho}}$$

where c is speed of sound, E is Young's modulus, and ρ is density.

Sound waves travel continuously through a medium, subject to absorption, until they encounter a medium with a different impedance. At this boundary, a portion of the wave will be reflected and a portion will continue to propagate through the new medium at a different speed and wavelength. If the difference in impedance is great enough, practically all of the wave will be reflected.

Acoustic impedance is the inherent resistance of a substance to a sound wave, defined as

$$Z = \rho \times v$$

where ρ is the density of the medium in kilograms per cubic meter, and v is the speed of sound through the medium in meters per second. Z is defined in units called Rayls. Air and lung have impedances less than 1 megaRayls, water 1.5 megaRayls, various soft tissues in the body range from 1 to 2 megaRayls, bone between 6 and 7 megaRayls, and lead zirconate titanate (PZT) ceramic approximately 34 megaRayls.

The absorption, or loss of energy, as a wave travels through a medium is given by the equation $\alpha = 1 - R_2$, where α represents the unitless absorption coefficient, and R represents the unitless reflection coefficient. The attenuation coefficient (decibels per centimeter at 1 MHz) of water is 0.002, blood 0.18, fat 0.63, muscle 1.3–3.3, and bone 5.

The sonography concept assumes sound travels in a straight line and meets perpendicular boundaries. However, when the boundary is oblique, the reflected wave will not return at a direct angle perpendicular to the oblique boundary.

Many logistics lead to non-ideal results in practice. An ideal boundary is smooth and uniform with clear demarcation between impedances. In many tissues, the boundary can be ill-defined. In addition, most of the layers of tissue the sound wave travels through are heterogeneous rather than a homogeneous material with a uniform impedance.

Components of a Sonography System

A sonography system is typically composed of a computer with a monitor linked to a probe that contains a piezoelectric transducer. Figure 9.1 shows

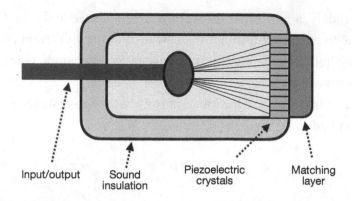

| Input/output | Sound insulation | Piezoelectric crystals | Matching layer |

Figure 9.1 Ultrasound probe. An ultrasound probe is made up of a set of piezoelectric ceramic elements connected to an electric power source that both provide charge to induce the creation of ultrasound waves and also carry back electrical signals generated from returning echo waves compressing the piezoelectric elements as they return. A backing or dampening layer behind the piezoelectric ceramic elements prevents retrograde transmission. A matching layer helps transition the wave to a resistance medium closer to the biological mediums through which it will pass. An acoustic lens helps shape the transmitted ultrasound waves to a desired focus. The probe is embedded in a sound-insulating material to limit waves to the forward-facing window.

a diagram of the probe. The probe is able to send out a focused ultrasound wave pulse as well as detect reflected waves, which are relayed back to the console for interpretation and translation into an image. Due to technological advancements, handheld devices, even phones, can be used in place of a workstation.

The heart of the ultrasound system within the probe is the piezoelectric element. As mentioned previously, naturally occurring crystals were the first substances found to have the potential to convert mechanical stress into electrical current, including quartz, topaz, and Rochelle salt. Over time, scientific efforts in industry worked to maximize this effect through ceramics to optimize the piezoelectric effect and durability in addition to minimizing size and weight. The most common ceramic used today is PZT. The material, in combination with its size and shape, will produce a particular natural frequency when electrically triggered to oscillate. This gives manufacturers control of the frequencies a particular probe may generate to correspond with desired imaging characteristics. To further increase the efficiency of the piezoelectric element, the PZT ceramic is built into the probe as a series of fine strips less than 1 mm thick, each coupled with electrodes. This is termed an *array*.

A backing material is used as a dampening layer to prevent retrograde ultrasound transmission and reduce inadvertent signal production (noise). This also permits finer control of produced vibrations within the ceramic, creating ultrasound pulses with shorter pulse width, which improves the resultant reflected wave signal-to-noise ratio.

Antegrade to the ceramic is an acoustic matching layer. The matching layer serves to gradually transition from the higher impedance of the PZT ceramic to the lower impedance of the body being examined. Without this intermediate medium, the difference in impedance would be such that there could not be efficient transmission into the object. The matching layer is typically made up of a series of resins with stepwise reduction in impedance.

The final component of the probe is the acoustic lens, which serves to focus the beam, control beam angle, and limit spread. Concentration of the ultrasound waves allows for increased transmission efficiency and thus return signal. Several types of material, especially variants of rubber, plastic, or silicone, can be used at manufacturer's preference. The lens should have an impedance that approximates the human body.

Gel is placed on the surface of the body where the probe will be placed in order to eliminate air between the two. Each probe can be adjusted to a range of pulse frequencies and duration, controlled by electric currents to the transducer.

Three basic types of probes are used in medical sonography:

- Linear probes are relatively higher frequency probes designed for higher resolution imaging of superficial structures and blood vessels.

- Curvilinear probes utilize lower frequency ultrasound waves to permit increased field depth at the reduced resolution, making this probe well suited for abdominal imaging.
- Phased array probes combine a restricted window with good field depth to facilitate upper torso imaging between ribs.

APPLICATIONS

One of the most well-known uses of medical ultrasound is evaluating a growing fetus in a pregnant woman. The lack of ionizing radiation permits safe imaging, and the rapid and continuous imaging allows for visualization of the active movements of the fetus. Visualization and measurements of the fetus are critical prenatal steps in assessing for proper neurologic development. Major congenital anomalies such as septo-optic dysplasia and holoprosencephaly can be detected before birth.

Ultrasound continues to be a valuable tool for neonates and infants. The soft fontanelles in newborns provide a convenient window for intracranial ultrasounds, giving the pediatric medical team a safe and readily available option to assess for acute damage, such as hemorrhage.

In adults, there are several essential uses for sonography in neurology. *Carotid duplex imaging* has been and continues to be a mainstay in stroke evaluations. Despite advances in CT and MR angiography, carotid assessment with ultrasound continues to be used in many circumstances. In patients who cannot receive CT iodinated contrast and have contraindications to MRI, ultrasound offers a safe form of vessel imaging with few contraindications. High-resolution static views offer reliable estimation of vessel stenosis. Doppler imaging gives functional data, allowing verification of partial blood flow versus complete occlusion. This information can be critical for decision-making regarding surgical interventions. Although most ideal for studying the carotid arteries close to the surface of the neck without overlaying bony structures, ultrasound can also offer partial assessments of the vertebral arteries.

The *echocardiogram* is another mainstay in post-stroke workup. Systolic function, atrial enlargement, the presence of patent foramen ovale, and detection of clots are evaluated for, helping identify intervenable risk factors for further stroke. Agitated saline injection during echocardiogram is commonly used in stroke patients to search for a right-to-left intracardiac shunt.

ARTIFACTS

Reverberation

Reverberation refers to an artifact created when the image field includes two highly reflective boundaries. As the distal-most echo wave returns toward the probe, it can be reflected back by the more proximal boundary and again when it reaches the more distal boundary a second time. This creates an artificial image effect of several equally spaced lines.

The probe itself can create a similar effect if a high-intensity echo is received from a reflective boundary such as metal. The echo can be partially reflected from the probe surface and return again. The second echo may be mistakenly interpreted as a deeper boundary given the additional time interval. This can repeat several times and create equidistant faux boundary lines known as a repetition artifact.

Mirror Image

A mirror artifact is a false redundant image of a real object created by an additional back-and-forth echo of the sound wave. As a sound wave is reflected from the source of a true object back to the transducer, it can then reflect back again from the transducer to that same object and return again. The difference in time is misinterpreted as a singular wave returning from a deeper object, so the duplicated false image appears in line and deeper to the true one. Figure 9.2 illustrates this artifact.

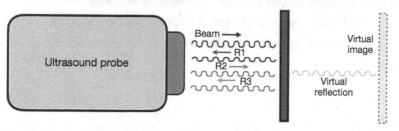

Ultrasound mirroring: first reflection (R1) returns to give correct response but part of the energy reflects back to the target (R2) and another echo returns (R3). The time for this travel creates a virtual image.

Figure 9.2 Ultrasound mirroring. Mirroring is an ultrasound artifact in which an echo wave can be reflected back from the transducer, meet a boundary a second time, and return once more to the transducer. The processing computer misinterprets a "back and forth" reflected wave as a single "out and back" wave that encountered a deeper boundary.

Acoustic Shadowing

In acoustic shadowing, the ultrasound beam encounters material that is either highly reflective or absorbent, obscuring the image distal to that material. Pockets of gas represent an area of high impedance change and create reflections or reverberations with a resultant poor signal-to-noise image beyond the material. This is called "dirty" shadowing due to the visual result.

In the case of a stiff, highly reflective object such as calcified tissue, insufficient ultrasound waves are able to propagate past the object, leading to a complete or near-complete absence of echo signal (anechoic). This is referred to as "clean" shadowing because the result is homogeneous black. Ultrasound operators can use the presence of this effect to help confirm the identity of abnormal tissues, such as renal stones.

A third form of shadowing is referred to as edge shadowing. This effect is seen if the peripheral components of the ultrasound beam encounter a curved boundary and are refracted away from rather than reflected back toward the probe. This also produces an anechoic resultant image.

Acoustic Enhancement

Acoustic enhancement can be thought of as the inverse occurrence of shadowing. In this situation, if a sound wave propagates through a tissue medium with low attenuation, a greater (less diminished) amplitude ultrasound wave will be present at the far boundary compared to other tissue mediums at the same depth. This results in a higher amplitude echo generated at the boundary and seen at the detector, as the return path will again be low attenuating and allow for a greater percentage of ultrasound waves to travel without absorption. The resultant image will show a brighter boundary than expected. This artifact can be advantageous in that its presence can suggest to an interpreter the type of tissue the path includes, such as a fluid-filled cyst rather than a solid homogeneous mass.

Beam Width

Sound waves naturally spread as they propagate. While the probe uses an acoustic lens to help narrow the transmitted signal, the beam will widen with depth. If the beam periphery enters a fluid-filled cavity, an echo signal can be created that falsely appears as heterogeneity within the fluid (pseudo-sediment).

Aliasing

This is an artifact specific to Doppler imaging that occurs when the sampling rate is not sufficient for accurate determination of flow direction and velocity. A visual example of aliasing is videos of spinning helicopter blades. When the blades spin at a faster rate than is captured by the video, the resultant images may appear to show slowly moving or still blades. We see exactly that while in the emergency department of our hospital watching the monitors while air ambulances land on the roof—the blades seem to be in slow motion while landing. In medical Doppler imaging, higher velocity areas of blood circulation can challenge the Doppler sampling frequency such that inaccurate findings result.

Speckle Artifact

Speckle artifact is cleverly named after its appearance, which comes from sound wave scatter from small structures within heterogeneous substances. The scattered waves lead to constructive and destructive interference with surrounding waves, seen as artifactually brighter or darker spots on the generated image.

Blooming Artifact

Blooming artifact is seen in Doppler ultrasound. With this artifact, the generated image displays an area of flow that is inaccurately large and beyond the actual boundaries of the moving substance. It is a "high gain" artifact that is a result of the inherently lower resolution of Doppler ultrasound relative to static grayscale ultrasound, and the effect is predominantly seen along the object boundary furthest from the transducer. It is also known as color bleed artifact. This is a key reason for the "duplex" aspect of carotid imaging: Doppler studies should not double as a method to obtain precise spatial information of the arteries.

CLINICAL CORRELATIONS

Carotid Stenosis: Diagnostic Carotid Duplex Ultrasound

Among the many conditions that increase the risk of stroke, carotid stenosis looms large in vascular neurology. Chronic vessel disease can involve both small and large arteries, and even partial blockage of one of the four great vessels providing blood flow to the brain has the potential for catastrophic consequences. Figure 9.3 shows a recording of a patient with

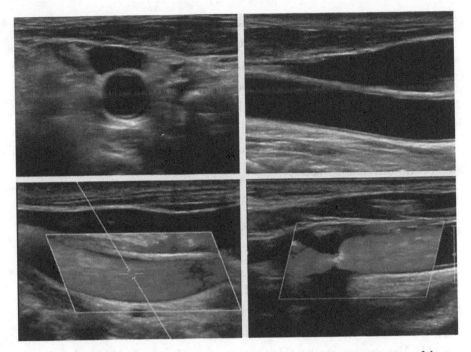

Figure 9.3 Carotid ultrasound of the carotid artery. (*Top left*) A transverse view of the carotid artery, in the middle of the screen. (*Top right*) A longitudinal section of the carotid. Above the carotid in these two figures is the jugular vein. (*Bottom*) Images show the Doppler results with normal flow in the section on the left and narrowed vessel with acceleration of flow velocity on the right, including a small amount of turbulence.

stenosis of the internal carotid artery producing some narrowing of signal and acceleration of flow velocity in the region of stenosis.

A common location for progressive plaque buildup is the area of carotid bifurcation, giving off the internal and external branches of the carotid artery. This branch point offers a point of turbulence and a natural location for plaque accumulation. Given its size, the carotids can accommodate 50% narrowing (stenosis) or more and continue to permit adequate perfusion if blood pressure and volume are normal. As discussed in fluid neurohydrodynamics , however, a critical stenosis or blockage may result in ischemic stroke and could be precipitated by hypotension. The anterior and middle cerebral arteries branch off of the terminus of each carotid artery, making loss of perfusion through here a harrowing prospect.

Certain stroke syndromes are suggestive of carotid stenosis, and initial contrast-enhanced CT imaging can often confirm this. In patients with renal impairment precluding iodinated contrast administration, or those needing more careful evaluation of the pre-cerebral arteries, carotid duplex imaging can be a valuable tool. The word duplex denotes that both structural and Doppler imaging make up the imaging assessment. Determination of peak systolic velocity is a key metric used in carotid intervention decision-making.

Subarachnoid Hemorrhage Vasospasm: Transcranial Doppler Ultrasound

Vasospasms are an abnormal focal constriction of arterial blood vessels that are a response to injury, vasculitis, or the local presence of blood byproducts as can be seen in the setting of a subarachnoid hemorrhage (SAH). Vasospasm can be a dangerous aftereffect in SAH because the restriction of blood flow can lead to an ischemic stroke. In many intensive care units, SAH patients are monitored for the development of vasospasms by use of ultrasound. Transcranial Doppler ultrasound is a technique used to evaluate the differences in flow velocities among major cerebral vessels in relation to the internal carotid artery. The portable non-invasive option for high-risk patient monitoring is also used for rapid intervention when warranted.

Essential Tremor: Therapeutic Ultrasound Ablation

Essential tremor, also termed benign or familial tremor, is a common movement disorder in which the patient develops a postural and action high-frequency variable amplitude tremor that is commonly in the arms, but it can involve the legs, head, or voice. It is often exacerbated by stress and can be alleviated by alcohol consumption. The disorder often has a clear inheritance pattern in families, but it can be sporadic as well. Several treatment options exist, including propranolol, primidone, topiramate, and gabapentin. However, these medications each have several potential side effects and contraindications that may limit their use in certain patients. Also, some patients may only have partial response to medications and some no appreciable benefit.

Deep brain stimulation can be effective for refractory tremor, as can ultrasound. Especially in patients with an asymmetric disabling tremor that is refractory to medicines, unilateral ultrasound ablation therapy may be warranted. As a non-invasive procedure, patients who otherwise may not be ideal surgical candidates can be considered. An MRI-guided focused ultrasound thalamotomy targets a particular region of the thalamus contralateral to the problematic limb, and a high concentration of high-powered ultrasound waves are transmitted, leading to controlled heating of the tissue. The limited area of tissue ablation offers the majority of patients significant reduction in tremor. Although generally well tolerated, it should be considered that this is an irreversible procedure and can have adverse effects even with successful targeting.

Anencephaly, Lissencephaly, Septo-Optic Dysplasia, and Spina Bifida: Prenatal Screening Ultrasound

The development of the nervous system in embryology is a fascinating topic that warrants its own book. It is an unfortunate reality that all the marvels that occur in biological development also have the potential to not occur successfully. A swath of congenital conditions exist, some preventable and others not. We discuss a few conditions that neurologists are well familiar with.

Anencephaly is a condition incompatible with life after birth in which the brain and skull have incompletely formed. This may be partial or the complete absence of brain tissue.

Lissencephaly is a condition of incomplete brain development, particularly the proper cortical development into gyri, resulting in agyria or pachygyria. Infants with this condition are born with microcephaly and can have several abnormalities, including failure to thrive and seizures.

Septo-optic dysplasia is a disorder involving the underdevelopment of brain midline structures, including the septum pellucidum, corpus callosum, pituitary, and optic nerves. The extent of underdevelopment, involved structures, and long-term deficits can vary greatly.

Encephalocele is the result of an incompletely closed neural tube, and thus skull, that can permit an outpouching of brain tissue. The survival rate is slightly greater than 50%.

Meningocele is a similar condition in which there is an incomplete enclosure of the spinal canal by the spine, allowing an outpouching of the meningeal coverings. Although spina bifida occulta is also an incomplete spinal enclosure, no components of the spinal canal protrude. By contrast, a myelomeningocele indicates the protrusion of both meninges as well as at least a portion of the spinal cord and/or nerves.

Prenatal ultrasound screening is an ideal tool for early assessment of fetal development because the developing skeleton does not represent the challenge that the dense adult cortical bone does, ultrasound does not transmit ionizing radiation, and ultrasound does not require prolonged motion suppression. In certain cases, conditions identified on prenatal ultrasound screening such as a myelomeningocele can prompt intervention measures, including prenatal surgery.

X-Ray and Computed Tomography

EVAN M. JOHNSON ■

FOUNDATIONS OF X-RAY AND COMPUTED TOMOGRAPHY

What Is X-Ray?

X-rays are a form of radiation that exist on the electromagnetic spectrum between ultraviolet light and gamma rays. X-rays are very high-energy, high-frequency photons with wavelengths in the range of 10 nanometers to about 100 picometers. As with all forms of radiation that travel the speed of light, frequency and wavelength have an inverse relationship. As described by the fundamental equation, the speed of light equals the wavelength times the frequency of the radiation.

In medicine, X-rays may be both diagnostic and therapeutic in their use. X-rays permit a form of basic body imaging and, in high targeted delivery, can destroy tumor beds to facilitate cancer therapy.

How Did X-Ray Imaging Come About?

X-ray radiation was first noticed incidentally in the late 1800s by scientists working with applications involving electron beams. These studies would unknowingly generate X-rays, effects from which were slowly realized by some of these physicists, including Philipp Lenard, Nikola Tesla, and Wilhelm Röntgen.

Röntgen is credited with the formal discovery of X-ray radiation for his work in 1895. His work involved generation of X-rays from cathode rays in a Crookes tube that was designed to block visible light but would still produce a glow on a fluorescent screen placed 1 m away. He further advanced this discovery by producing images of his wife's hand using a photographic plate, creating a silhouette of her bones and ring, and began to realize its medical potential for non-invasive imaging of internal structures. He also gave the radiation its famous identification as X, as he wanted to refer to it as an unknown source. He received a Nobel Prize in physics for this discovery.

The scientific community quickly realized the potential use for X-rays in clinical scenarios that next year. Only 1 year later, the first evidence of X-ray radiation poisoning was realized.

What Sets X-Ray Apart from Other Imaging Modalities?

X-rays represent some of the oldest means of imaging the human body non-invasively. Extremely small wavelengths allow for incredibly high anatomic resolution to be captured.

In medical usage, X-rays are relatively inexpensive, quick in their generation, and allow for rapid screening and assessments, such as imaging the lungs and abdomen. Video fluoroscopy is an important tool for proceduralists, who can perform their work with real-time non-invasive image guidance, allowing procedures such as heart catheterization and cerebral angiograms to be performed. X-rays also form the foundation on which computed tomography (CT) is based. As previously mentioned, radiotherapy helps ablate tissues such as tumor masses.

The primary disadvantage of X-ray radiation is that it is ionizing in nature and therefore a carcinogen. Ionizing radiation indicates that X-rays can impart enough energy to electrons that they escape their atomic or molecular orbit, which results in the generation of ions. Ionization is harmful from a health perspective, in that ions are inherently less stable and can subsequently break chemical bonds and/or form highly reactive free radicals. The formation of free radicals in particular can lead to DNA damage and thereby increase the risk of inducing cancer by way of genetic mutation.

How Does X-Ray Work?

The basic principle behind X-ray generation in modern medical devices is that a very high-energy electron beam is created and targeted from a cathode emitter to an anode target, commonly tungsten. As the electron beam encounters the dense anode target, changes in the electron beam velocity result in changes in the electron beam energy that are realized in the form of heat and X-ray generation.

What Is CT?

Computed tomography is a diagnostic imaging technique that utilizes X-ray projections across 360 degrees to produce non-invasive cross-sectional imaging of objects.

Computed tomography is not the perfect name for this technique anymore because it actually refers to the processing of multiple projections to create a cross-sectional result, a technique that today applies to many forms of diagnostic medical imaging, including positron emission tomography (PET) and single-photon emission CT. CT scans are often referred to as CAT scans, which stands for computerized axial tomography scan. Again, this is not specific to X-ray–based tomographic imaging.

How Did CT Come About?

The mathematical theory that made CT possible, known as the Radon transform, was developed by Austrian mathematician Johann Radon in 1917. In 1961, William Oldendorf, a neurologist at the University of California, Los Angeles, built a prototype scanner utilizing an X-ray source opposite a detector which he rotated around an object to generate a series of radiographs that he would combine to produce cross-sectional images. He first imaged a nail surrounded by nails, which would not allow any single angle of X-ray to capture the center nail in the middle of the outer ring of nails. He received a U.S. patent for his work in 1963, and the first guidelines for diagnosis of medical pathology were established in 1968, primarily for abdominal use. EMI produced the first commercial CT scanner in 1971. It has been said that the revenue generated by The Beatles in the 1960s provided the funds for which EMI developed its CT scanners, although it is unclear if it is truly the case. The first EMI scanner was dedicated to producing images of the head, generating 80 × 80 pixel cross sections in approximately 11 minutes.

The next generation of CT scanners was developed at Georgetown University employing higher grade technology in the detector and computer systems and was acquired by drug company Pfizer, which helped begin the national spread of CT as a more common medical imaging tool in hospitals. Even today, evolving technology continues to transform the capabilities of CT scanners, allowing for improved resolution, reduced radiation burden, and increased throughput.

What Sets CT Apart from Other Imaging Modalities?

Computed tomography scans share many of the strengths and weaknesses of X-ray imaging, providing high-resolution non-invasive images at the cost of ionized radiation administration. The ionizing radiation dose of CT scans far exceeds that of a single X-ray image. CT scans are also more sensitive than single X-ray projections to motion artifact due to increased acquisition times. If a percentage of radiographs within a series are not well aligned with the remainder, the resultant cross-sectional composite will be of poor quality.

How Does CT Work?

A CT scanner is most simply described as the gantry that holds an X-ray source opposite a digitized detector allowing for 360 degrees of rotation around the object to generate a series of X-ray projections that can then be used to produce tomographic reconstruction and a series of cross-sectional images.

The gantries were originally designed for step and shoot imaging, in which they would rotate a small degree and then pause for an X-ray projection. Today, the gantries allow for continuous motion and rapid X-ray projection acquisition without stopping.

The full details involving tomographic reconstruction are not explored here. However, some important aspects of image reconstruction are discussed.

To allow for image normalization, it is important that scanners be calibrated. A detector acquisition without X-rays being sent, sometimes referred to as a *dark*, allows for the normalization of pixels inappropriately generating signal in the absence of X-rays and establishing one set of values for the maximum spectrum of the scanner. A second detector acquisition with X-rays being sent without an object present, referred to as a *bright* scan, allows for normalization of dead pixels that are not registering signal despite X-ray interaction with the detector element, and it also establishes the maximum signal of the detector spectrum.

Standardization of CT values allows for comparison scans to be studied over time and at different facilities. To do so, CT scans are produced to conform with a grayscale known as *Hounsfield units* (HU). This scale, named after its creator Sir Godfrey Hounsfield, is specific for CT. With proper calibration and normalization, CT scans will produce a grayscale such that air will have a value of –1,000 and water 0. Most other tissue types, such as fat, soft tissue, and bone, do not have a singular value and do not have true fixed units on the scale.

Generally, CT scanners utilizing X-ray beams generated from electron beams accelerated at the same power source voltage will produce very similar values for these tissue types. The power source voltage is described in units of kVp, whereas the energy of the electron beam and its subsequent X-ray photons is described using keV. kVp stands for peak kilovoltage, the maximum voltage used to power the acceleration of the electron beam. The energy of the beam and the X-rays cannot exceed this. keV denotes electron volts, the kinetic energy of the electrons traveling with the accelerated beam.

Because CT scans can be produced with different energy spectrum of X-rays, the value of tissues can vary somewhat despite calibration of HU. For example, dense cortical bone is usually the highest attenuating tissue in the human body. If a CT scan is produced using lower energy X-rays, those X-rays are more likely to be attenuated in bone than if a CT scan was acquired using higher energy photons with greater penetrating power. Because of this, a CT created by lower energy (KeV) X-rays versus one created with higher energy X-rays will both have an air value of –1,000 and a water value of 0, but they will have different values of the same cortical bone. The lower energy X-rays will be more attenuated and result in a higher HU compared to the high-energy X-ray scan. This difference is not linear and does not demonstrate change equally for different molecules. The change in attenuation in accordance to change in X-ray energy is the attenuation coefficient, which is specific for each type of material. This property provides the fundamental principle behind dual-energy CT scanning, which utilizes the specific change in attenuation across energy to distinguish specific types of substances, such as blood versus iodinated contrast. Related, each element has a characteristic K-edge—a significant shift in the attenuation coefficient curve corresponding with photon energy near the binding energy of the innermost electrons of the element (K-shell).

WHAT ARE THE COMPONENTS OF AN X-RAY AND CT SYSTEM?

X-ray generators are composed of high-voltage electrical source in an X-ray tube consisting of a heated cathode source, vacuum tube, and anode target, typically tungsten.

Electron Beam

The electron beam is generated by heating a cathode source powered by a high-voltage power source. The voltage powering the X-ray tube should be referred to as kVp (peak kilovoltage), rather than simply kV (kilovolts). The energy of the electrons produced by the source, keV (kilo-electron volt), is limited by the kVp but can exist at lower energy.

The cathode source is typically a filament made of a metal such as tungsten. As the high-voltage electric current is sent across the filament, it heats to a degree at which electrons obtain enough (kinetic) energy that they leave their atomic orbits and can be propelled toward the anode source.

The electron beam is directed toward the anode target through the vacuum with magnetic fields. In higher powered electron beam generators, a magnetic maze can be created to continuously accelerate the electron beam and amplify the energy generated.

X-Ray Generation

As the electrons are released and propelled toward the anode target as a directed beam, the real magic happens. Typically, but not always, a tungsten anode serves as the wall that the electron beam will slam into and release its energy in the form of heat and radiation.

As the electrons encounter the highly positive nucleus core of atoms in the anode target, one of several possibilities may take place. The inner target can be thought of as a dense field of atoms for the electrons to encounter. Although it is possible for some electrons to fully pass through the anode, the majority will interact with the tungsten atoms in their path. A direct hit is possible, but probabilities make it more likely that the electrons will come near enough to an atomic nucleus that they will change velocity due to attraction between the negatively charged electrons and positively charged atomic nucleus. Coulomb's law can be thought of as an atomic analog to gravity in space. Just as a comet passing close to a sun will alter its path due to gravity, electrons passing near an atomic nucleus will likewise change direction as that electrostatic force acts upon them. This recalls another fundamental principle of physics, which is that energy may be changed but cannot be added or destroyed. By changing the electron's velocity, of which direction is a component, the energy of the electron affected by an atomic nucleus must also change. To account for this change in energy, energy is released by the electron. This is referred to as braking radiation or the bremsstrahlung effect.

In truth, the vast majority, greater than 99.9%, of the energy produced by the electron beam encountering the tungsten target is represented by heat. Less than 1% is converted into X-rays.

Also, it is important to again discuss probability. Although it is common for X-rays to be referred to as being generated as a specific kilovoltage, the produced X-rays are in fact a spectrum of energies, which cannot exceed that of the electron beam from which they were created. The produced energy can be low, such as the heat that was previously mentioned, or high enough to exist on the X-ray spectrum. The bulk of the X-rays generated from a tube will typically be one-third to two-thirds that of the energy of the electron beam from which they were generated. The spectrum of energy will be specific to the X-ray tube used.

There will also be the characteristic K-line (K-edge) that is representative of the particular anode material used. The K-edge is a phenomenon of photon attenuation that reflects an element's unique K-shell binding energy. The K-shell is the innermost orbit of electrons in an atom. Although all K-shells may only hold up to two electrons, the binding energy is more varied between elements because the positive charge of the nucleus and the repulsive outer electrons are significant contributing factors. With regard to radiography, elements with a K-edge that matches the energy of a photon have a potential to attenuate those X-rays to a much higher degree. This phenomenon is exploited by contrast agents such as barium and calcium, which have K-edges (33 and 37 keV, respectively) that match with the lower end of the generated energy spectrum of X-rays. By altering the kVp, one can directly affect the resultant keV of the produced X-ray beam.

Because of this wide spectrum of X-rays generated by the source that are not necessarily desired for imaging purposes, filters such as aluminum or copper are often used in X-ray systems to limit the energies of X-rays sent from the source to the detector. This helps limit the amount of beam hardening seen in the resultant image, a phenomenon that is discussed later as a CT artifact. In advanced systems, multiple interchangeable filters can permit the same system to utilize beams of different energies.

Also, because X-ray generation from the electron beam can be in all possible directions, a "window" is made so that only X-rays created in the direction of the detector are allowed to travel outside of the source. This window can further be shaped depending on the type of X-ray beam desired, allowing for slit, fan, or cone beams to be produced. A filter placed at the site of the window will help by removing lower energy X-rays from the spectrum created.

Of note, a phenomenon that is more important to recognize in CT than in plain radiography is the heating of the anode target.

As mentioned previously, the vast majority of energy created in an X-ray tube is heat. This means that the tungsten anode target is subject to significant heating while in use. If you recall, heating of a metal will result in expansion of that metal. Expansion of the metal can therefore shift, if only slightly, the center of the X-ray beam. When generating an image based on 360 degrees of exposures, a shift in the center of the beam can result in the series of exposures not being fully in line with each other. This potential source of artifact is typically handled by appropriately warming the scanner and bringing the tungsten anode up to its working temperature and volume. X-ray generators are often built with a rotating tungsten anode to help disperse and distribute the heat more equally, like a rotisserie chicken.

Detector

Opposite the X-ray source is a detector, and between the two is the object being imaged. Figure 10.1 shows a diagram of a CT scan detector.

As the X-ray travels to the detector, one of three events may occur. The X-ray may have sufficient energy and a path such that it travels unadulterated and strikes the detector with the same energy it had when it left the source. The X-ray could also be attenuated, meaning that its energy could be absorbed or significantly altered by atoms that it encounters as it passes through an object. This can include air, which should be distinguished from a vacuum. An attenuated photon will not reach the detector. A third possibility is that an X-ray may have its path altered as it encounters particles along the path, which is referred to as Compton scatter effect.

A detector will therefore produce an image not terribly unlike a shadow puppet. X-rays, similar to light, will either be able to travel uninterrupted to the detector or be attenuated. Recall that X-rays with a wide spectrum of energy are generated. Because of this, X-rays at very different energies can pass through the same path, encountering similar objects but resulting in different events. Higher energy X-rays are more likely to pass through objects, whereas lower energy X-rays are more likely to be attenuated.

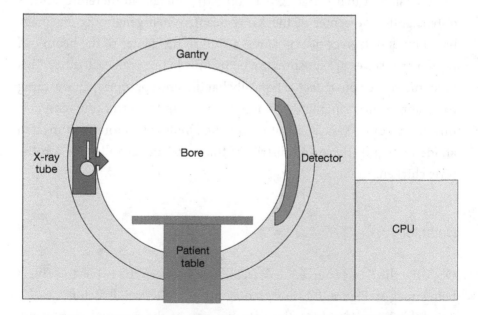

Figure 10.1 CT scanner. Diagram of a CT scanner with source and detector surrounding the table. CPU, central processing unit.

Objects that lie in the path between the source and the detector also affect the likelihood of attenuation. An atomic coefficient refers to a molecule's likelihood of attenuating an X-ray. Just as you can imagine air with spaced out gaseous molecules has little likelihood of attenuating an X-ray, dense metals such as lead or copper have a very high likelihood of attenuating an X-ray. The human body has multiple types of tissue with very different capacities for attenuation, making X-ray imaging useful. Our lungs, especially when we take in a deep breath, are very low attenuating given their high gas content. Our bones, specifically our dense cortical bones composed of calcium, are a high-attenuating tissue. Fat is lower attenuating by nature than more protein-rich organs such as muscle. It is because of these different properties that the X-rays generated for medical imaging provide more useful information than a simple shadow.

A couple of takeaways from the previous discussion are that as a spectrum of X-rays pass through an attenuating object, such as bone, fewer X-rays will reach the detector than were sent from the source, and the energies of the X-rays that reach the detector will generally represent those higher along the spectrum.

The proximity of the object related to the source and detector will result in the magnification of the image. If you imagine the X-ray beam exiting the window of the source as light and the detector as a wall, you can picture using your hand to make a shadow puppet. The closer your hand is to the light source, the larger the shadow puppet will appear on the wall. If you were to place your hand close to the wall, the shadow puppet would be much smaller and similar in size to your hand.

Originally, X-ray radiographs were produced on film, where the energy carried by the X-rays would strike an emulsion consisting of a material such as silver halide and produce silver ions, which would clump and opacify the area in which the X-ray struck.

In the modern era of electronic medical information, charge-coupled detectors have largely supplanted the original film-based technique. In these detectors, small filters ensure that only direct X-ray beams may strike the detector itself, minimizing noise from scatter. The detector converts the energy of the X-rays striking into an electric signal that is digitized.

One basic method of translating a beam of X-rays into a digital signal is photo-stimulated luminescence. The plate absorbs the energy of the projected X-rays, and that energy is then transformed into a luminescent signal after stimulation by visible light such as a laser. Photomultipliers then collect, amplify, and send the signal to an analog-to-digital converter.

Indirect detectors also utilize this basic principle of transforming X-rays first into light before translating them into an electrical signal. Indirect detectors utilize scintillation to make this work. X-rays first strike a layer of a scintillator, typically gadolinium- or cesium-based, which converts the beam of X-rays into light. This layer is directly attached to a silicon detector array composed of extremely small pixels, each with a thin film transistor that contains a photodiode to translate the light from the scintillator into an electrical signal that is tied to the specific detector element.

There also now exists direct X-ray to signal conversion that allows the X-ray photons to be transformed into electrical charge without a middleman. A photo conducting substance such as selenium absorbs and converts the photons into an electric charge. The energy of the photons is relayed by the generated current. The direct conversion without a visible light intermediary improves the sharpness and resolution of the detector.

Detectors may be built of varying sizes and shapes. For simple radiography, a shape that corresponds with the size of the object of interest is created. Earlier Fan Beam CT scanners utilize a linear detector that incorporates a very restricted width, resembling a slit. Flat-panel detectors are utilized in some more modern scanners, allowing for a larger field to be imaged with each pass of the CT scanner.

The modern rate of evolution in detector technology allows manufacturers to produce new models of scanners that may create high-quality images with lower amounts of radiation or images so rapidly obtained that they can capture the structure of the beating heart without degradation due to the motion.

Gantry

The gantry is the ring that secures all of the components of the CT scanner and spins rapidly around the object of interest. One of the more breathtaking moments in my life was being inside a room as an uncovered CT gantry was brought up to full speed. Although that may sound underwhelming at first, it becomes more awesome as you consider that these systems are spinning components with the combined weight of a steel car at 100–200 revolutions per minute (i.e., two or three rotations per second). It is impossible not to imagine the incredible damage that would take place if all components of the scanner were not perfectly secured to the gantry, given the centripetal force being generated.

WHAT ARE SOME OF THE COMMON USES OF X-RAYS AND CT IN NEUROLOGY?

Cerebral Angiography

Cerebral angiography is an advanced imaging technique to evaluate the vasculature supplying the brain. This can be helpful in situations in which vasculitis or aneurysms are suspected.

Angiogram is performed under fluoroscopy, which can be thought of as X-ray–based video monitoring. A catheter is place through a patient's accessible artery and directed to an artery of choice, such as the carotid artery, where, with live X-ray imaging, iodinated contrast dye is injected into the vessel and the filling of the downstream vasculature may be observed and recorded in real time. This allows observations to be made of the vessel structures, looking for stenosis or other abnormalities, as well as functional assessment, observing the fill rate. In cases of suspected vasculitis, a calcium channel blocker may be injected in an area of stenosis where a vasospasm is suspected. If the injection of the calcium channel blocker in the area of stenosis results in vessel dilation, it can be presumed that the vessel narrowing was due to spasm and not a plaque.

Many neurosurgical techniques and procedures rely on cerebral angiography to direct the procedure.

Non-Contrasted Head CT

Sometimes referred to as the extra vital sign in the emergency department, a non-contrasted CT scan of the head can often be the first diagnostic procedure for patients encountered by neurology. This is a rapid way of assessing the basic anatomic structure of the skull and brain. It does not offer as much tissue detail as magnetic resonance imaging (MRI) of the brain might provide, however, but is rapid and will demonstrate gross abnormalities.

CT imaging excels at differentiating between tissues of different radiopacity, particularly air, fat, water, protein, and bone. It is common to hear people say that blood is bright on CT, which is only partially correct. What is "bright" or "dark" is a relative concept dependent on the window (absolute range of HU values in the grayscale) and level (middle value point; lower HU values will be darker and higher HU values will be brighter). When looking at scans, you will notice that the vasculature of a normal person does not stand out against soft tissue. You will not be able to appreciate the four chambers of the heart without contrast enhancement, for example. However, acute and subacute bleeds do stand out with noticeably higher HUs than the surrounding soft tissue, although still noticeably lower in value than even slightly calcified bone. The difference lies in clotting. Blood clots extrude fluid and concentrate hemoglobin within the clot. Hemoglobin, specifically its iron content, is radio-opaque. It takes the body many weeks to fully absorb all of the blood products, just as a bruise can take some time to fully heal, meaning that a bleed on the brain can be apparent for some time after the actual event.

In a normal person, the non-contrasted head CT readily shows the skull and sinuses, orbits, the parenchymal brain tissue, the falx, and the ventricles.

The usefulness of a non-contrasted head CT is manyfold. In the setting of an acute change in neurological status, a non-contrast head CT can help rapidly rule in or out several chief concerns on the differential diagnosis. An intracranial hemorrhage, be it a subdural, epidural, subarachnoid, or intraparenchymal event, should be apparent. Figure 10.2 shows a bright area that is an intracranial hemorrhage in the thalamus.

Abnormally hypodense areas are also a potential critical finding. Most commonly, this finding indicates vasogenic edema, which can represent one of many abnormal processes, including infection, malignancy, or a fully established stroke at least 6 hours old. Gray matter (higher protein content, hence slightly brighter than white matter) and white matter (higher lipid content due to myelin, hence slightly darker then gray matter) are relatively well differentiated in a normal CT scan, but this distinction can be lost as one of the earlier findings in a stroke.

Ventriculomegaly is another common pathological finding easily seen on plain CT. Enlarged ventricles can represent one of many abnormal processes, such as communicating or obstructive hydrocephalus. One must carefully consider the clinical context because atrophy of the brain may also naturally lead to expanded ventricles, a benign finding referred to as hydrocephalus ex vacuo.

CT Angiography

The next most common scan performed in neurological CT imaging is CT angiography (CTA). For this, an iodinated contrast agent is injected intravenously with a scan timed such that the arteries of the patient should be saturated with the contrast agent but not the venous system—in other words, acquisition of images during the arterial phase of the injection. This permits the provider to carefully assess the arteries supplying the brain. In settings of a suspected acute stroke, this may demonstrate a severe stenosis or a complete occlusion that may explain the vascular event. This also can provide neurosurgeons information as to whether a clot has formed in an area that is accessible for potential thrombectomy.

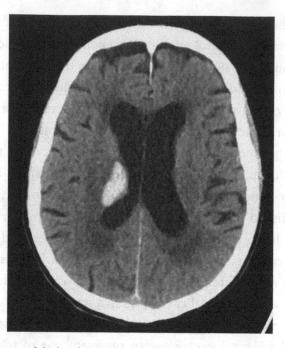

Figure 10.2 CT scan of the head. CT of the brain that shows a hyperdense lesion in the right thalamus. This is an acute hemorrhage.

Figure 10.3 shows a CTA of a patient with occlusion of the right middle cerebral artery with reconstitution of some flow downstream.

If both a CT head and a CTA were obtained preserving patient position, a subtraction technique can be applied and produce images that only display the vasculature itself, representing the only difference in values between the two scans. This can be helpful in many instances because it subtracts nonvascular tissue and allows for a more focused observation of the blood vessels.

CT Perfusion

The perfusion scan builds upon the CTA technique, but it is more advanced. Rather than producing a static structural image of the vasculature, functional data are obtained. This can be thought of as a volumetric analog to the previously described fluoroscopy technique. By continuously imaging the head while contrast is delivered, a CT perfusion scan provides information including the mean transit time and mean volume delivered to areas of brain tissue. This can inform the provider of which areas are receiving normal blood flow and in which areas blood flow is impaired or completely cut off. In the case of strokes, this is critical information that informs to the penumbra, or endangered but salvageable tissue, and the core of the infarct, or tissue that is already irreversibly lost. An image showing a large penumbra with a small core suggests that there is a potential for an improved outcome if the impeded blood flow can be restored, such as in the setting of a successful thrombectomy.

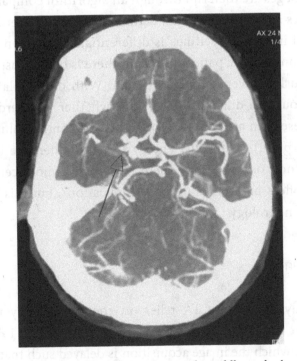

Figure 10.3 CT angiogram. CT angiogram shows right middle cerebral occlusion as indicated by arrow. Flow is reconstituted by collaterals distal to the occlusion. Compare with the opposite side, which is normal.

Dual-Energy CT

We previously discussed that different materials have inherently individual attenuation coefficients, meaning that as the energy of an X-ray changes, so will the corresponding attenuation and resultant HU (except in the case of air or water). Soft tissue, bone, and iodine, for example, do not change values at a set percentage change compared to one another when imaged at different energy CT scans. This unique property allows for the differentiation of similar but different materials using dual-energy CT.

The basis of this technique is to image an object twice, once at a relatively low-energy CT (but still adequate to fully penetrate the object and not result in detector starvation) and again at a higher energy. The resultant images are then run through an algorithm comparing the different values against known attenuation coefficient. Where this can be most useful in medical settings is differentiating between blood from iodinated contrast in a patient in whom there is both a suspected bleed and previously delivered contrast media. With only a plain CT technique, it would be difficult to ascertain whether a hyperdense signal might represent an area of blood or an area of contrast blush (residual tissue absorption). With dual-energy CT, the difference between iodine and blood or iron can be determined with more certainty. This technique offers multiple additional applications, but this is the most common in neurology.

CT Venogram

As previously discussed, CTA relies on contrast delivery followed by a subsequent scan in the arterial phase. A CT venogram is an imaging technique in which the image acquisition is delayed such that the venous system is maximally contrasted and arterial phase contrast is minimized.

This technique is important in neurology for evaluation of the venous sinus system, particularly to evaluate for obstruction such as a venous sinus thrombosis (VST). VSTs are easily missed on plain CT and even CTAs, and they can be life-threatening if not treated.

CT Myelogram

A myelogram is a contrast-enhanced image of the spine. In some cases, MRI of the spine is unobtainable, such as in patients who have a permanently implanted pacemaker that is not compatible with MRI.

A myelogram is obtained by performing a lumbar puncture and injecting contrast material through this site, providing contrast between the spinal fluid and spinal cord. This can be helpful in cases in which structural abnormalities such as an abscess or traumatic cord injury are suspected.

CT-PET

Positron emission tomography scans are discussed further in Chapter 12. Most modern PET images are coupled with a CT scan to provide anatomical mapping. PET only displays signal from positron emission radiation produced from a radiotracer and otherwise does not provide high-quality anatomical information. Therefore, CT scans are often obtained at the same time and fused to the PET image so that the provider has appropriate mapping to determine the location of areas of significant tracer uptake.

CONTRAST AGENTS

Barium

Barium sulfate is the most commonly used contrast agent for X-ray–guided gastrointestinal (GI) studies. Like other heavy metals, barium is high attenuating, therefore radio-opaque, and provides excellent soft tissue contrast when lining the GI tract. Barium in particular is preferred because it is also not readily absorbed in the GI tract and is generally nontoxic. Note that should there be a bowel perforation and barium enters the peritoneal space, it can cause significant irritation and peritonitis or mediastinitis.

As an aside, few patients, if any, remark at the great taste of barium-based oral contrast agents. After taking the taste test challenge myself, I personally recommend banana flavor if available.

Other non-barium oral solutions exist for the imaging of the GI tract. In fact, if a patient were to ingest an adequate amount of chocolate milk, the calcium in the drink would also provide contrast enhancement. However, people who are sufficiently concerned about the state of their intestines that they want to image them should probably not be drinking large volumes of dairy.

Iodine

Contrast-enhanced imaging techniques of CT typically rely on iodinated material. Metal elements naturally provide high radiopacity, a property not exclusive to iodine. In fact, gold provides the highest attenuation of common materials available. Copper, lead, iron, aluminum, and most other metals provide sufficient capacity, even at low concentrations, to produce contrast enhancement relative to soft tissue.

The choice of material to use in medical imaging comes down to tolerance as well as cost. Most metals are highly toxic to the human body in greater than trace amounts. The symptoms of Wilson disease, for example, result from toxic copper buildup in the body. Iodine cannot be directly injected in its elemental form. Commercially available formulations are created to be relatively non-absorbable, non-ionic, and renally excreted. The typical molecular form of these molecules is a central ring with three iodine groups and three larger complex chains that reduce the possibility of interaction of the three iodine molecules with other substances. Even so, allergic reactions to iodine-based contrast are fairly commonplace and can range in severity from skin flushing and vomiting to severe renal toxicity and even anaphylactic shock.

ARTIFACTS

Motion

X-ray imaging is rapid but relies on the patient to be still while the image is being acquired. Motion while the exposure is taking place will result in blurring and loss of anatomical accuracy. Respiratory motion in particular can be problematic, so patients are directed to either hold a deep breath or fully exhale for the image, depending on the interest of the radiographer.

Like many other medical imaging modalities, CT scans are sensitive to patient motion. CT exposure times are very brief, on the order of milliseconds or less, but substantial patient motion during the course of the scan can lead to significant image degradation.

Even if a person were to stay still, respiratory and cardiac motion are more difficult to control. Patients are frequently asked to take in a deep breath and hold it because that will suppress respiratory motion and a scan can often be acquired in the time it takes to hold a single breath. Also, there is the added benefit of filling the lungs with air, which provides contrast against soft tissue structures. Deep inspiration will also expand the lung and provide space between the tissues.

Cardiac motion cannot be suppressed voluntarily, but some specialized modern CT scanners offer cardiac imaging. These scanners continuously acquire ultra-brief exposure times, simultaneously monitor the patient's electrocardiogram, and then later collate exposures based on their timing within the cardiac cycle. A series of CT scans can be produced, each representing a different phase of the cardiac cycle. The result is a pristine image of the heart without motion degradation, as well as four-dimensional functional information of the heart beating.

Partial Volume Averaging

Partial volume averaging is an artifact seen in most digitized images. As opposed to film images, digital images are composed of pixels, or voxels in the case of three-dimensional reconstructions.

Consider that the object of interest is scanned by a field of X-ray waves that are each smaller than any detector element, which are then translated to a pixel that can only hold a single value. Technology limits how small a pixel can be created, and each pixel that is generated is composed of multiple X-rays that could otherwise represent even finer resolution. Each pixel is made up of multiple values but can ultimately only represent a single value, which will be an average of all the signals received. This is known as partial volume averaging.

When a pixel overlays the border between two different materials or tissue types, that border will be visualized in a stepwise manner. If an object much smaller than a single pixel is imaged, its value will be averaged with surrounding tissue.

As modern scanners provide increasingly greater resolution, this artifact becomes less of an issue clinically.

Streak Artifact

A phenomenon known as beam hardening takes place as a polychromatic X-ray beam passes through objects of high attenuation. Lower energy X-rays within the beam are predominantly attenuated, leaving the higher energy X-rays as the beam exits the object; this is referred to as beam hardening.

When objects are imaged with a substance that has a very high attenuation coefficient, such as metal, beam hardening takes place and commonly results in several forms of artifact, including streak. If you image a person with considerable dental work, the resultant scan in the same plane as the metal will be degraded in quality and with streak artifact.

The streak represents a misinterpretation of the received photons at the detector. When a very high attenuating object such as metal almost completely absorbs the X-ray beam, detector starvation may occur.

Streaking may also occur to a lesser degree between two opaque objects. This is one challenging aspect of imaging the brain, which is surrounded by a dense skull.

Metallic elements are generally the highest attenuating objects encountered in X-ray imaging. If a dense metal object, such as a surgically implanted hip replacement, is imaged, the high attenuation of the metal and its capacity to cause streaking, scatter, and detector starvation may be such that it obscures objects in the same plane or in very close proximity.

Cupping

Cupping is an additional artifact related to beam hardening. If you consider that CT scans are generated by imaging across 360 degrees of an object, different degrees of beam hardening will take place if a beam passes directly through the center of a dense object versus angles where the beam may only tangentially cross the object.

Because of these differences, the values in the center of a high-density object may often be lower than those along the periphery. This phenomenon has been recognized for some time, and many scanners have software correction for this built-in.

Ring

Ring artifacts represent a detector element failure that has not been appropriately compensated for with calibration.

PHYSICS EQUATIONS RELEVANT TO X-RAYS AND CT

The most fundamental equation in circuitry may be the interdependence of voltage, current, and resistance. Voltage supplies the electromotive force that compels electrons to travel from a state of high energy to a lower energy state. Resistance, measured in ohms, is the system's inherent oppositional properties that counter electron flow:

$$V = I \times R$$

Voltage = Current × Resistance

The power (P) delivered by a system is a function of both current and voltage. As current is determined through voltage and resistance, the delivered power can also be calculated as the square of the voltage divided by the resistance:

$$P = I \times V$$

$$\text{Power} = \text{Current} \times \text{Voltage}$$

$$P = \frac{V^2}{R}$$

Power = Voltage squared divided by resistance

The energy of photons is related to their energy and wavelength:

$$E = \frac{hc}{\lambda}$$

Energy of an X-ray = (Planck's constant × speed of light) divided by wavelength

Simplified, the energy of an X-ray is inversely proportional to its wavelength, as the other two values in the equation are fixed:

$$C = f\lambda$$

In photons, the speed of light equals the frequency times the wavelength.

Because the speed of light is a fixed value, the wavelength of a photon determines its frequency and vice versa:

$$KE = eV$$

The kinetic energy of an electron beam = electric charge of an electron × voltage

The charge of an electron = $1.60217662 \times 10^{-19}$ coulombs

The following is Coulomb's law of force of attraction between charged particles:

$$F = k_e \frac{q_1 q_2}{r^2}$$

where k_e is Coulomb's constant ($k_e \approx 9 \times 109 \ \text{N·m}^2 \text{·C}^{-2}$).

In physics and chemistry, the law of conservation of energy states that the total energy of an isolated system remains constant; it is said to be conserved over time. This law means that energy can neither be created nor destroyed; rather, it can only be transformed or transferred from one form to another.

Magnification is a useful tool in medical imaging, helping maximize the resolution of objects within the body that may be otherwise so small as to not be readily perceivable to the eye. The position of the object relative to the source and detector determines the level of magnification that takes place.

$$M = \frac{H_i}{H_0} = -\frac{D_i}{D_0}$$

where M is the magnification, H_i is the height of the image, H_0 is the height of the object, D_i is the distance from the lens to the image, and D_0 is the distance of the object to the lens.

$$F = ma$$

Force = mass × acceleration

$$F_c = \frac{mv^2}{r}$$

Centripetal force = mass × centripetal acceleration = mass × velocity squared divided by the radius

Attenuation coefficient measures the number of downward e-foldings of the original intensity that will be had as the energy passes through a unit (e.g., 1 m) thickness of material so that an attenuation coefficient of 1 m^{-1} means that after passing through 1 m, the radiation will be reduced by a factor of e, and for material with a coefficient of 2 m^{-1}, it will be reduced twice by e, or e^2.

CLINICAL CORRELATIONS

Acute Ischemic Stroke: CT Head, CTA Head and Neck, and CT Perfusion

Ischemic stroke, in its simplest definition, is irreversible brain damage caused by nontraumatic lack of oxygen relative to metabolic demand.

Most commonly, this is due to an interruption of arterial blood supply. The most common causes of this are embolic clots that form elsewhere in the body and lodge within an artery delivering blood to the brain or a local thrombus resulting from long-term vessel disease and plaque buildup. In either case, the tissue downstream from the blockage will be at risk unless there is adequate collateral blood supply or the clot is resolved. Acute stroke interventions are possible, especially at certified stroke centers throughout the country, if patients are assessed within a few hours of the onset of symptoms. Tissue plasminogen activators can be administered if within 3–4½ hours and contraindications are ruled out.

A non-contrasted CT head (CTH) is therefore an essential part of the acute stroke assessment in the emergency department. This scan may be obtained in less than 5 minutes and permit the provider to quickly review for various findings to inform the decision to administer or not administer the tissue plasminogen activator. Evidence of hemorrhage is a contraindication and will drastically alter management. Hypodense areas could represent a number of things, including complete infarction of tissue, tumor, or abscess. Any of these findings would be a contraindication because they would each carry an increased risk for subsequent intracranial hemorrhage. A formal assessment for completed middle cerebral artery stroke burden, termed the Alberta Stroke Program Early CT Score (ASPECTS), is based on evaluation of a series of distinct regions of the affected side of the brain for hypodensity. This offers an objective scale to help in the interpretation of the CTH scan when making medical decisions for management of the stroke patient.

Following the more time-sensitive decision regarding administration of a "clot buster," or in cases in which patients are assessed outside of the time window for that medicine, an additional CT scan of the head and neck including iodinated contrast agent is performed—CTA. The importance of this scan is that it provides a rapid screening of the larger vessels providing blood flow to the brain. Blockages can be identified in arteries as small as some distal branches of the middle and posterior cerebral arteries. Areas accessible via mechanical thrombectomy (up to a proximal M2 branch, as an example) can be visualized, making CTA a valuable part of the stroke assessment to determine if there is a large vessel obstruction that could be amenable to neurosurgical acute intervention.

A final phase of CT imaging in an acute stroke assessment setting might be CT perfusion, in medical centers with this capability. In perfusion imaging, the patient is continuously scanned while a bolus of contrast is administered. This permits serial monitoring of the various regions of the brain and quantified data of how much and how quickly iodinated contrast (as a surrogate marker for oxygenated blood) is taken up in discrete areas of the brain. Key data points, including mean transit time and time to peak uptake (T_{max}), can be compiled for assessment of brain perfusion. Areas of viable but endangered tissue receiving inadequate blood flow (penumbra) can be identified, as can areas of completed infarct (core). This information can further inform the decision of whether to attempt neurosurgical intervention.

Venous Sinus Thrombosis: CT Venogram

The human body is an incredibly complex machine, accommodating a continuous interplay of biochemical reactions maintaining homeostasis essential to life. Among these many biochemical reactions that maintain a carefully balanced equilibrium are the body's clotting and thrombolytic factors. This balance permits reaction to blood vessel injury without allowing excess clotting that can threaten proper blood flow. Just as conditions such as hemophilia result in a system imbalance in which clotting factor deficiencies impair proper arrest of bleeding, other conditions can tip the balance toward abnormal excess clot formation: thrombophilia, or a hypercoagulable state. This could be an inherited condition such as factor V Leiden, an indirect consequence of an infection or inflammatory disorders such as lupus or Behçet's disease, or other states including pregnancy and dehydration.

Cerebral venous sinus thrombosis (VST) is a potential consequence of a hypercoagulable state, in which an occlusion or partially obstructive clot forms and impedes proper venous drainage of the brain's blood supply. If the flow of a venous sinus is arrested, the forward flow of arteries feeding into that tissue bed is threatened, and this can result in ischemic stroke. Sequela of VST can include seizure, neurological deficits, and increased intracranial pressure, in addition to ischemic as well as hemorrhagic strokes. Early recognition is important because this gives the best chance to lyse the clot with blood thinners. VST can initially be occult, or only present with mild signs or symptoms that could go unrecognized. CT venogram (CTV) offers an accurate assessment of the cerebral venous system and is both a sensitive and a specific test for this condition. MR venograms are an additional option, although the specific flow dynamics of a partially occluded vessel may affect the accuracy of the finding compared to a CTV (see discussion of time of flight imaging in fluid dynamics and MRI in Chapters 4 and 11).

Normal Pressure Hydrocephalus: Non-Contrast CT Head

Normal pressure hydrocephalus (NPH) is a form of non-obstructive hydrocephalus without a clear identified cause for the excessive buildup of CSF causing ventricular dilation. Over time, this expansion of the lateral ventricles in particular can apply a stress on neighboring brain tissue, usually including the corona radiata. Patients may develop insidious and progressive deficits in their walking, micturition, and cognition. Specifically, people with NPH can develop an apraxic "magnetic" gait in which they have difficulty initiating steps, with it appearing as if their feet are stuck to the floor. Cognitive processing can be slowed, requiring more time and effort to perform tasks such as math or puzzles while generally sparing semantic memory. A neurogenic urinary continence is the third common symptom making up the notorious NPH triad. Note that not all three symptoms may be present at the time of diagnosis, and they tend to develop in different stages from one another. Although the definitive diagnostic test is a large-volume lumbar puncture demonstrating an objective improvement in cognition and/or ambulation, a non-contrast CTH can be a viable screening tool to suggest the condition. The triad of NPH symptoms is notoriously nonspecific and not uncommon to be found in the elderly population. Specific features on a CTH may suggest the possibility of NPH as the underlying reason for the symptoms. Ventriculomegaly is not in and of itself suggestive of NPH because it may be seen in several conditions. An increased ratio greater than 0.3 of the diameter of the frontal horns of the lateral ventricles to the maximal internal diameter of the skull (Evans' index), widening of the temporal horns of the lateral ventricles, an upward bowing of the corpus callosum, dilated sylvian fissures, and a narrowing of the cingulate sulcus relative to the anterior half are all features that may prompt further clinical investigation in the presence of a neurological exam consistent with NPH.

Glioblastoma Multiforme: CT Head with and Without Contrast

Glioblastoma multiforme (GBM) is the most common primary brain tumor in adults. It is an aggressive grade IV form of astrocytoma, the classification of cancer originating from astrocytes (glial cells rather than neurons). It is recognized as a rapidly progressive, malignant tumor with a propensity to cause both local intracranial hemorrhage and systemic coagulopathy, including pulmonary embolism. It is so aggressive in its replication and expansion that core necrosis is common; the tumor outstrips its ability to draw in nutrients to maintain its extremely high metabolic demand despite causing local angiogenesis. Given its rapid growth, extensive surrounding edema is common. Patients may present with symptoms of elevated intracranial pressure (severe headache, nausea, and vomiting) and/or neurologic deficits, personality change, or seizures. Initial assessment including rapid CTH imaging, with and without contrast, can often lead to early identification, prompting comprehensive workup and likely biopsy to confirm the diagnosis via pathology and confirm if the tumor may be responsive to chemotherapies. A methylated O^6-methylguanine-DNA methyltransferase promoter indicates responsiveness to temozolomide.

Certain features on CTH imaging are suggestive that a brain lesion may be a GBM, Including irregular borders, core hypodense necrosis without significant calcification, areas of hemorrhage, extensive surrounding edema, and an irregular pattern of enhancement. Although many tumors "respect the midline," that is not the case with GBM, and if seen, it may be another suggestive feature on CT imaging.

Magnetic Resonance Imaging

EAMON DOYLE AND EVAN M. JOHNSON ∎

HISTORY AND DEVELOPMENT OF MAGNETIC RESONANCE IMAGING

What does it take to create magnetic resonance imaging (MRI)? Just four Nobel Prizes, sneaking around late at night disassembling university property, and a clam.

Nuclear Magnetic Resonance

Since its development in the 1970s, MRI has become a key diagnostic technology for clinicians and researchers in every specialty of medicine. Although MRI can seem impossibly complicated to many newcomers, it is rooted in a technology first discovered in 1938: nuclear magnetic resonance (NMR). After Otto Stern's 1922 discovery of quantum spin states of protons, Richard Ernst discovered that the nuclei of certain atoms behave like tiny magnets. In 1950, Erwin Hahn discovered that when radio frequency (RF) energy was applied to these nuclei, they would temporarily absorb the energy and then release it back in a measurable way. Let's get into some of the details.

Main Field and Precession

The topic at the heart of MRI is *nuclear magnetic resonance*. In fact, MRI is more appropriately called "nuclear magnetic resonance imaging," but the word "nuclear" has a tendency to freak people out (unnecessarily, in this particular case). NMR is the property whereby certain atoms will spin around an externally applied magnetic field. In order to generate any signal via NMR, we must start with two things: a bunch of atoms that have a non-zero "nuclear spin" and a very strong magnet. The nuclear spin comes from the number of protons and neutrons. An odd number of neutrons and odd numbers of protons each provide a spin of ½: If the numbers of both protons and neutrons are odd, the spin is 1; if one is odd, the spin is ½; and if both are even, the spin is 0. In MRI, we do not generally need to speak of nuclear spin, so moving forward, the term "spin" is used interchangeably to refer to any nuclei that can generate an NMR signal. When any of these spins are placed in a powerful magnetic field (referred to in NMR and MRI as the "main field" or the "B_0 field," or simply B_0 or B_{naught}), they will align its magnetization vector with that field, spinning around the magnetic field like the Earth spins around its axis. If the magnetization vectors are knocked out of alignment with this field, they will begin to rotate around the field, a phenomenon called "precession." Furthermore, they will try to return to equilibrium with the magnetic field, a process called "nuation."

If you read other books on MRI, most will tell you that the spins are either "parallel" or "anti-parallel" with the field. This is likely a misinterpretation of the idea that a spin can have either an "up" state or a "down" state. This is not strictly true unless you can measure exactly one spin, which is not practically possible. Suffice to say, in a magnetic field, all the spin distributions will add to form a magnetization vector along the main field, and that magnetization vector is proportional to the main field and the spin density. This nuance has no practical effect on understanding or applying principles of MRI, so there is no need to argue with your engineering professor or attending radiologist over it even if they are wrong (but here is a citation if you are feeling feisty: https://doi.org/10.1002/cmr.a.20123).

Free Induction Decay and Echoes, or How We Form
Signals—How We Form Signals—How We Form
Signals—How We Form Signals . . .

In any case, you find yourself in a lab with a sample of spins that you have placed in a powerful magnetic field. If you wish to receive any signal at all from this sample, you need to find a way to disturb the equilibrium of these spins by knocking them out of alignment with the main field. This is accomplished by using an RF pulse—a burst of energy in the form of electromagnetic waves oscillating at a frequency tuned to the sample. This frequency is termed the Larmour frequency, and it can be determined by multiplying the gyromagnetic ratio of the nuclei species (e.g., hydrogen-1 or carbon-13) by the main field strength. By sending an RF pulse at this frequency, the spins will be tipped out of alignment with the main field. As they return to equilibrium, they will create their own electromagnetic signals oscillating at the Larmour frequency that can be received.

When the energy is simply received in this manner, it is called a free induction decay (FID), and this data can be processed to yield NMR spectra. If an RF pulse at time = 0 is followed by another RF pulse at time = t, energy will be released at time = $2t$, a phenomenon called a *spin echo*. All the data acquired with an NMR are based on these two methods of gathering signal, although they can be encoded in a variety of ways that will change the contrast in the resulting image.

The release of energy follows an exponential decay process described by three time constants, T_1, T_2, and T_2^*. Figure 11.1 illustrates graphically this discussion:

- T_1 describes how long it takes for the "longitudinal" or Z component of the magnetization vector to return to normal after excitation. It follows the equation $Mz(t) = 1 - M_0\, e^{-t/T_1}$,

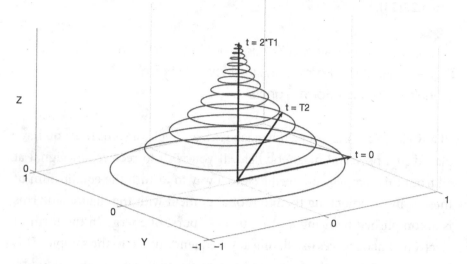

Figure 11.1 $T_1/T_2/T_2^*$ recovery/decay. This figure demonstrates the process of T_2 decay, T_1 recovery, and precession after an RF excitation pulse is applied. If a 90• pulse is applied, all of the magnetization is tipped into the transverse plane. The magnetization then spins around the B_0 field. The magnetization vector appears to shrink and then grow again because T_1 recovery takes longer than T_2 decay.

where M_0 is the initial longitudinal magnetization, and t is time after the initial excitation. T_1 is the longest of the constants and theoretically bounds T_2. T_1 cannot be directly measured because it does not affect the $X-Y$ component of the magnetization (called the "transverse" magnetization), which is the portion of the magnetization that can be measured.

- T_2 describes the decay time of the transverse magnetization, or the magnetization in the $X-Y$ plane that can be detected. As energy is lost, the $X-Y$ component of the magnetization vector gets shorter, causing a weaker signal. By performing multiple NMR spin echo experiments with increasing echo times, the maximum echo amplitudes will plot an exponential decay curve described by the equation $M_{xy}(t) = M_{xy}(0) \, e^{-t/T2}$.
- T_2^* describes how long it takes for the excited spins to lose phase coherence or to not be pointing in the same direction. This

causes the apparent loss of signal that is seen for a FID. The FID following a single excitation will decay via $M_{xy}(t) = M_{xy}(0)\, e^{-t/T2^*}$. It is theoretically bounded by T_2: It is always shorter than T_2, although if the magnetic field were perfectly uniform and there were no magnetic inhomogeneities to cause the spins to lose coherence, T_2^* would be measured as equal to T_2. It is important to note that T_2^* is descriptive of the scanning environment rather than the medium.

Note that magnetization is not conserved over the course of transverse decay and longitudinal recovery. This means that the magnetization vector is not required to stay the same length. In other words, if T_2 decay is extremely fast and T_1 recovery is extremely long, the magnetization will seem to disappear in the X–Y plane while not reappearing along the Z axis until much later.

MRI PHYSICS AND ENGINEERING

If you have gotten this far, you have covered three Nobel Prizes' worth of discoveries. That's pretty impressive; give yourself a pat on the back. We're sure you are wondering how you can take those squiggly looking NMR spectra and turn them into pictures of your insides. Using the principles of NMR to perform imaging requires some simple (although not necessarily easy) refinements and additions to the NMR machine used. Paul Lauterbur devised the theory that is the basis for MRI in the early 1970s. In the dead of night, he disassembled and modified an NMR machine to act like an MRI, performed experiments, and then removed his modifications so the machine would function as expected by the rest of the Stony Brook University chemistry department. In this manner, he successfully imaged vials of water and a clam.

Despite the initial rejection of publication by *Nature*, Lauterbur was successful in publishing his findings. Peter Mansfield, a physicist, began to work on improving MRI techniques, and the two were able to practically demonstrate imaging to the point that it was subsequently developed for medical imaging purposes. They were awarded a Nobel Prize for their troubles in 2003.

But how did they actually do it?

Before we start designing our perfect MRI machine, we have to choose what we are imaging. No, we don't mean a clam; we mean what kind of atoms. Although NMR works for any nuclei with a non-zero spin, every species of nuclei has a different Larmour frequency. The antennas used to transmit and receive RF need to be tuned to that frequency and the RF pulses have to be modulated at that frequency (more on that later) in order to excite the spins or receive any signal back from them. Therefore, most MRI systems are designed for just one species. In the clinic, MRI is universally associated with hydrogen. It's in water. It's in fat. It's attached to everything in your body. In fact, hydrogen makes up approximately 9.5% of human bodies by mass and 62% of the total number of atoms. If oxygen or carbon exhibited NMR, we'd image them instead, but alas, we're stuck with hydrogen.

Extending NMR to Imaging

B_0 FIELD UNIFORMITY

Now that we know what we are imaging, we can start designing the machine. The first major change we need to make compared to an NMR spectrometer is to make the B_0 field extremely uniform. The reason for this is simple: We want all the hydrogen spinning at the same frequency. In order to form an image with MRI, we need to be able to encode location information into the signal; we accomplish this by subtly changing the frequencies of spins at different locations in the MRI. Because the Larmour frequency is field strength dependent, each hydrogen nucleus will only have the same precession frequency if it experiences the same magnitude field. Thus, the main magnetic field in MRI machines is designed to vary less than one part per million (equal to 0.0001%) over the whole imaging bore.

GRADIENT FIELDS

Now that all the protons are precessing at the same frequency, we will do our best to make them spin at different frequencies but in predictable ways. To accomplish this, a new set of three magnetic fields are added to the MRI called gradients. Unlike the main magnetic field, which always remains at the same strength, gradient fields can be changed over time. When a gradient is activated, it makes a linear change in the magnetic field. Thus, on one side of the bore, the field will be slightly weaker and on the other side it will be slightly stronger. This means that everything will spin slightly slower on one side that on the other for as long as the gradient field is turned on. This allows us to achieve spatial encoding.

The addition of the gradient fields also allows us to generate a new type of echo, the gradient echo.

REFERENCE FRAMES

Now that we have gradient fields, that gives us the concept of direction. In order to determine where in space a particular group of protons is, we need a reference frame. In MRI, there are two reference frames that matter: the lab frame and the rotating frame. The lab frame is fixed in physical space relative to the MRI machine; the z axis points along the B_0 field, whereas the x and y axes are perpendicular to the bore. The z axis of the rotating frame is also parallel with B_0, but the other axes, x' and y', spin around B_0 at the Larmour frequency. Thus, if you imagine yourself in the rotating frame, the protons would not seem to be spinning; rather, the room would.

K-SPACE

This bit is not strictly a change to the MRI machine, but it is nonetheless useful here to begin understanding how to think about data acquisition. When we speak of k-space, we are referring to the numerical space in which we acquire data. These are the raw data that are transformed into an image later on. Keep this in the back of your mind for when we discuss data processing. For the time being, just think of k-space as the relationship between the gradient waveforms and the data.

THE MRI MACHINE

Modern MRI machines are complex, but their basic elements can be represented as shown in Figure 11.2.

The Basic Anatomy of a Pulse Sequence

OK, we've done it. We've converted our NMR machine into an MRI. But how do we actually use it? Before we get into exactly how RF pulses, spatial encoding, data acquisition, and post-processing work together to form images, we need to get a little terminology out of the way. When we play an RF pulse and a series of gradient waveforms and then read data back from the scanner, that is called a pulse sequence. A simple pulse sequence diagram is shown in Figure 11.3. This image shows a basic gradient echo pulse sequence. This is a great first sequence to get to know because it is

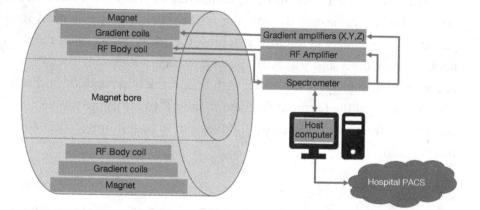

Figure 11.2 Schematic of an MRI system. This is a simple block diagram of an MRI system. The magnet bore contains the superconducting magnet coils, the gradient coils, and the RF coils. The host computer communicates with the spectrometer to specify the pulse sequence, and the spectrometer then sends signals to the RF and gradient amplifiers to play the pulse sequence. Data are then received from the RF coil, converted to a digital signal by the spectrometer, and the raw data are sent back to the host for reconstruction. After the image is reconstructed, it can be viewed or transmitted to the hospital picture archiving and communications system (PACS).

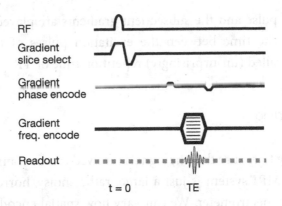

Figure 11.3 Pulse sequence diagram. This figure demonstrates a simple gradient echo pulse sequence with a slice-selective RF pulse. The slice selection gradient plays at the same time as the RF pulse. The phase encoding gradient then selects the line in *k*-space to be read. Finally, the frequency encoding gradient is active while the readout occurs, encoding the location of the samples along the readout line as the gradient echo occurs.

the basis for many of the most commonly used sequences. Furthermore, it demonstrates the three main spatial encodings.

Moving from top to bottom (and essentially left to right) in Figure 11.3, the first things we see are the RF pulse (top line) and slice selection gradient. The RF pulse performs excitation of the tissue, whereas the slice selection gradient ensures that only a slice of tissue is excited. Slice selection is not a requirement for MRI, but it leads to substantial scan time reduction and is therefore included in the vast majority of clinical imaging sequences. After excitation, we have the phase encoding gradient. This selects the *k*-space line that will be acquired. Due to the Fourier transform of *k*-space to form our image, there is not an easily comprehended link between a line in *k*-space and a line in the image; every *k*-space sample contributes to every image pixel. Finally, we see the readout gradient. This, approximately speaking, selects which pixel in the *k*-space line is being sampled while the analog-to-digital converter is reading the signal and converting it to bits from the computer to record.

This, however, is not the complete pulse sequence. This combination of RF pulses and gradient fields will be repeated multiple times with slight spatial encoding variations to acquire all of *k*-space. Each time an excitation RF pulse and the subsequent gradients are played, it is called a repetition. The time between the excitation pulses of two adjacent repetitions is called (unsurprisingly) repetition time or TR.

Spatial Encoding

Without a way to ensure that the signals received can be transformed into an image, an MRI system is just a large, rather noisy, horrendously expensive NMR spectrometer. We can vary how spatial encoding happens by changing the timing, strength, duration, and shape of the gradient fields and the RF pulses. How we do this has a major impact on how our resulting images look.

PHASE ENCODING

After excitation, all of the spins will be aligned in the same direction—a direction that is not parallel with the B_0 field. By applying a phase encoding gradient, the angle of the vector with respect to the x axis will change for the spins depending on where they are in the magnet bore. We can differentiate spins along the phase encoding direction by how much phase they have. This essentially allows one line in k-space to be selected at a time. So, a given pulse sequence is played repeatedly, and the only change each time is the strength of the phase encoding gradient. Once all of the lines in k-space have been acquired, the scan is finished.

FREQUENCY ENCODING (THE READOUT GRADIENT)

When we receive the signal, it will be a sum of waves at various frequencies. If we turn on the gradient field while we are reading data, hydrogen on one side of bore will spin slightly faster than hydrogen on the other side of the bore; the signals will therefore oscillate faster or slower depending on where in the bore they are. By separating the waves by their frequency, we know which side of the bore they came from. This encodes data within each selected k-space line. Together with the phase encoding gradient, this allows us fill our two-dimensional (2D; or even 3D) k-space, which will later be transformed into an image.

SLICE SELECTION

When an RF pulse is applied that excites the entire 3D volume of interest, known as a "nonselective" or "hard" pulse, the resulting signal is a sum of all of the excited atoms, requiring phase encoding in two directions. Such an approach is slow, causing very long scan times. If a gradient field is played during an appropriately shaped RF pulse, the set of spins that are excited can be limited to a single slice. Thus, we can essentially select a single plane (or slice) of tissue to be excited, removing the need for one of the phase encoding directions, which vastly decreases scan duration.

Radio Frequency

The RF pulse is key for excitation. Its amplitude and duration are directly proportional to the flip angle that is achieved during excitation; the flip angle is the angle created between the $+Z$ axis about which the protons spin and the magnetization vector after excitation. So, a 90 flip angle would place the magnetization in the X–Y plane. This will change image contrast as well as the amount of energy that is deposited in the patient. The amount of energy deposited through scanning is called the specific energy dose (SED), and the rate over time that it occurs is called the specific absorption rate (SAR). Certain limits are imposed for safety reasons, and some implanted medical devices have their own manufacturer-tested limits that must be respected to prevent device malfunction or tissue heating and burns.

In addition to the amplitude, the "bandwidth" of the pulse is an important quantity for slice-selective RF pulses. For a specific slice selection gradient amplitude, a higher or wider bandwidth will encode a thicker slice, whereas a lower or more narrow bandwidth encodes a more narrow slice.

Signal-to-Noise Ratio

The signal-to-noise ratio (SNR) is the ratio between the magnitude of the signal produced by a given tissue and the magnitude of the background noise that is present. If you activate the receiver on the MRI without doing any excitation, you will notice that a small amount of random signal is recorded anyway. This "thermal" noise is always present and is a result of imperfections throughout the MRI hardware. Engineers go to great lengths to reduce this noise. Practically speaking, the SNR is improved when large flip angles are used because transverse signal is increased. In addition, having a large acquisition window (i.e., turning the receiver on for a longer duration during the pulse sequence) also improves the SNR because more data can be acquired. Noise can also be introduced by patient movement such as breathing or cardiac motion. Such noise is called "physiologic" noise, and reducing it requires slightly different strategies.

Contrast Mechanisms

Due to the complex way in which MRI signals are formed, there are a number of techniques that can be used to generate contrast between tissues in different ways. They range from changing pulse sequence parameters to introducing exogenous contrast agents that cause certain tissues to enhance relative to others.

T_1

T_1 contrast is generated when two tissues have widely differing T_1 values. Using a TR that is much shorter than one of those T_1 values prevents the longitudinal recovery of that species of spin between excitations. This has the effect of reducing the longitudinal magnetization vector, which then reduces the amount of magnetization that is tipped into the transverse plan on each excitation, causing the long-T_1 species to appear dark and the short-T_1 species to appear bright.

Gadolinium is a metal that has a strong shortening effect on T_1. It is generally injected into the venous system, causing the T_1 of blood to drop and appear brighter on T_1 images. For exams looking at blood vessels, such as MR angiography, gadolinium improves the SNR. When assessing tumors, the vascularization of the tumor can become more apparent by comparing pre- and post-contrast scans. Furthermore, tissue perfusion can be assessed by performing rapid scans as gadolinium is injected and watching it "wash" into tissues.

T_2

Similarly, T_2 contrast is generated when two species have differing T_2 values. When an echo time is selected that is long relative to the shorter T_2, the transverse signal from that species will decay more before acquisition, causing it to appear darker on the resulting image.

T_2^*

T_2^* contrast is generated when there is a disturbance in the magnetic field that causes the spins to dephase, leading to an apparent reduction in the transverse magnetization. This type of contrast is indicative of the environment that the protons are in. Thus, T_2^* is always shorter than T_2.

Iron is another metal that demonstrates a strong effect on MR images. Due to its ferromagnetic properties, it reduces T_2 and T_2^*. Thus, it causes iron-loaded tissues to become darker. This property is leveraged in the brain to assess traumatic injury because iron will accumulate after blood vessel damage.

STEADY-STATE FREE PRECESSION

A special type of scan called steady-state free precession is created when the TR is shorter than both the T_1 and T_2 of the imaged species. This causes echoes to form immediately before each RF pulse. The resulting contrast is dependent on T_1, T_2, TE, TR, and flip angle where TE is Time to Echo and TR is Repatition Time. Notably, it shows contrast between tissues with differing ratios of T_2 and T_1, making it valuable for differentiating tissues such as blood and myocardium. It also is very fast, opening up applications in flow imaging.

DATA ACQUISITION AND ANALYSIS

Coils

Every MRI system has at least one RF coil that is used to transmit RF pulses and receive data. It is built into the main housing of the MRI and is called the *body coil*. It generally has two elements that can be used to capture the varying field created by the spinning magnetization vectors, providing both signal amplitude and phase. In addition, most systems have additional coils that can be plugged into image body parts specifically. These coils are usually "receive-only," although certain head coils also perform RF transmit to improve the uniformity of brain imaging or reduce the SAR and SED without degrading image quality. Newer systems even have coil elements built into the table and select elements closest to the region of interest to improve the resulting images.

k-Space and Fourier Transforms

The Fourier transform is the underpinning mathematical relationship between the data acquired by the MRI system and the images that are produced. If you are not a mathematician, prepare yourself.

Imagine that the data we have acquired are described by a function, $\mathcal{F}(\omega)$, where ω is a point in our "frequency" space. This should at least seem viable to the newly inducted MR enthusiast reading this because the gradients cause each point in the acquired *k*-space to be related to frequency.

Now, consider that our image is described by a function $f(x)$, where x refers to a particular location in the image, just like ω referred to a position in k-space. We can convert our k-space data into an image by performing an inverse Fourier transform:

$$F(x) = \int\limits_{-\infty}^{\omega} F(\omega)e^{2\pi i x \omega}d\omega$$

OK, that probably doesn't mean much to the average reader, so let's walk through it. In English, this would be read as "$f(x)$ equals the integral from negative infinity omega to infinity omega of $F(w)$ times e raised to the exponent of 2 pi time i times x times omega." Great, that was helpful. Put more plainly, you can think of k-space as ingredients in a smoothie recipe, the image as the smoothie, and the inverse Fourier transform as the blender. For every location x in our image, you can calculate the value of that pixel by multiplying k-space by a function that relates frequency and location and then adding it all together. Thus, every k-space sample contributes to every image pixel.

The previous equation is technically the 1D form of the inverse Fourier transform; in practice, we take the 2D or 3D inverse Fourier transform because we need to perform the transform for both x and y dimensions, as well as the z direction if the acquisition used two phase encoding directions rather than a slice-selective RF pulse. This is mathematically equivalent to performing it once along one axis and then doing it again along the other axes.

Encoding

In k-space, two types of encoding can be performed: frequency and phase. In a traditional Cartesian k-space trajectory, the k-space is acquired in lines. The phase encoding gradient "selects" the line, with a stronger gradient selecting a line further up or down along the y axis and a weaker gradient selecting one closer to the origin. This gradient is played after the excitation but before the acquisition. If a nonselective RF pulse is used, a second phase encoding gradient will be played to encode along the z

axis. The frequency encoding gradient is then played during acquisition, causing the resonant frequency of each point in space to be slightly different, encoding along the x axis.

APPLICATIONS OF MRI

Types of Sequences

GRADIENT ECHO

In the world of MRI, the most simple pulse sequence imaginable is a gradient echo. In this pulse sequence, there is a single RF excitation pulse followed by one or two phase encoding gradients and then a frequency encoding gradient. The frequency encoding gradient first dephases the spins and then reverses, causing an echo to form as the spins rephase. These pulse sequences are simple, tend to have low SAR/SED, and show strong T_2 and T_2^* contrast.

SPIN ECHO

A pulse sequence with an excitation RF pulse followed by a 180 inversion pulse (technically, an RF pulse) will generate a spin echo. This pulse sequence is robust to spatial variations in magnetic field strength, making it very useful in cases in which magnetic susceptibility causes signal loss and also in cases in which T_2^* contrast should be minimized.

This sequence can be accelerated by placing multiple 180 pulses sequentially following the excitation pulse. Such sequences are known as fast spin echo or turbo spin echo sequences and can reduce imaging time by reducing the number of repetitions required to completely fill k-space.

INVERSION RECOVERY

Inversion recovery sequences are unique because they take advantage of the T_1 recovery of a selected tissue to nullify the signal. It can be thought of like a gradient echo or spin echo sequence with an additional 180 RF pulse that is played prior to the excitation pulse. This causes all of the spins to be inverted completely, at which point their longitudinal magnetization begins to recover from $-Z$ to $+Z$. The time between the pre-pulse and the excitation pulse is tuned so that the magnetization vector of the nulled species is just reaching a length of 0 as the excitation pulse is played. At this point, the 0-length magnetization species cannot be excited by the RF pulse, so they generate no signal while species with a non-zero longitudinal magnetization do produce signal.

A special case of this is the *fluid attenuated inversion recovery* (FLAIR) scan, a popular neuroimaging technique. The pulse sequence is a T_2-weighted spin echo sequence with the inversion pre-pulse timed to nullify cerebrospinal fluid (CSF). This results in an image that shows T_2 contrast but without the presence of CSF, allowing for the identification of lesions that may be filled with CSF or another fluid, such as blood.

SATURATION

In MRI, there is a broad category of sequences that use saturation techniques to change the tissue contrast. One of the most common is "fat saturation," which takes advantage of the different frequency shifts of fat relative to water to nullify fat. Other chemical saturation techniques perform similar functions but at different target frequencies. At their core, all of the saturation techniques operate by designing an RF pulse to target spins at a particular frequency. With no spatial selection gradient, all spins that experience the RF pulse will be excited, as is the case with fat saturation.

However, if a gradient is played during the saturation pulse, a plane of spins can be saturated, as is the case with a spatial saturation band. Spatial saturation can help prevent flow or other artifacts from confounding the acquired images.

SPECTROSCOPY

Because MRI is in essence a complicated NMR spectrometer, it stands to reason that we should be able to just do spectroscopy, and this is in fact the case. When imaging is performed, a gradient is played during readout to encode the data spatially in one direction. By instead performing phase encoding along the axis that is normally the readout gradient (for a total of two phase encode gradients for slice-selective excitation or three if a nonselective pulse is used), data that are not encoded by a gradient can be read from a single voxel at a time. The recovered data, once an inverse Fourier transform is applied, are the spectra of that particular voxel. This is used in neuroimaging to create a chemical profile of the brain to investigate metabolites, ischemia, and other phenomena.

Advanced Topics

FUNCTIONAL MRI

Functional MRI (fMRI) technically refers to any MRI images that demonstrate function of a system—for example, a knee flexing under load. However, colloquially, fMRI has come to refer to a brain imaging technique that attempts to determine which parts of the brain are active when the subject is presented with some sort of stimulus. This is usually accomplished via a technique called BOLD for blood oxygen level-dependent contrast. When neurons are more active, they require more oxygen, so oxygen in blood becomes more depleted in active areas than in non-active areas. Soon thereafter, the blood flow to that brain region increases correspondingly. The deoxygenated hemoglobin is paramagnetic (enhances magnetic field), whereas the oxygenated hemoglobin is diamagnetic (resists magnetic fields). The enhanced magnetic field that occurs in the active regions causes reductions in T_2 and T_2^*, which then lead to temporary reductions in signal intensity. The signal difference is small, only approximately 1% different from baseline. By repeatedly taking images with and without stimulus, statistical techniques are used to identify regions where the signal decreased.

DIFFUSION

Diffusion-weighted imaging (DWI) is important for a number of diagnostic techniques, including stroke and tumor identification. Diffusion imaging works by playing sets of bipolar gradients (i.e., gradients that first are positive, and then reversed, so that the total area under the gradient set is 0) after the initial RF pulse but before the encoding gradient. Any spins that are stationary will dephase and then rephase perfectly because they will have experienced equal but opposite gradients. If a spin is moving, it will not rephase completely due to the spatial gradient dependence, causing the signal from that proton to decrease. When the image is encoded, spins that demonstrated less diffusion will produce more signal than those that did move, allowing for a map of diffusion to be created.

Diffusion maps will demonstrate increased diffusion in areas where cell membranes are breaking down and water can move in a less restricted manner. Because the speed of proton motion can vary, scanners with DWI sequences allow for the adjustment of the "b-value," which specifies the strength of the diffusion encoding gradients. Spins that generate signal at high b-values have low diffusion coefficients.

A single diffusion-weighted image cannot by itself demonstrate the diffusion coefficient. For this, we need two images, one with a b-value of 0 and one with a non-zero b-value, to calculate the apparent diffusion coefficient (ADC). The ADC is computed from the b-value (b), the signal intensity at the b-value (S), and the signal intensity at a b-value of 0 (S0) as follows:

$$ADC = \frac{1}{b} * -\ln\left(\frac{S}{S0}\right)$$

By extending DWI to multiple directions, the directional path of water molecules can be interrogated. Whereas DWI uses gradients in one direction, diffusion tensor imaging uses diffusion gradients in many directions—as few as 3 but often up to 64 or 128 directions—to determine the diffusion coefficients for each direction at every location in the image. By combining these individual DWI images, the diffusion tracts can demonstrate the shapes and paths of groups of neurons.

MRI ARTIFACTS

As with any complicated system, the data you receive are not always what they seem. For a variety of reasons, MR images can sometimes show features that are incorrect, warped, or completely fictional. Any image feature in MRI that does not correspond to an underlying physical reality is called an artifact. Let's discuss some of the more common MRI artifacts.

Fat Shifts

The use of Fourier transforms to convert k-space data to images leads to some interesting artifacts if there are tissues with differing center frequencies being imaged at the same time. A very common artifact is called a fat shift. It occurs in gradient echo images because protons attached to fat demonstrate a spectral shift in their NMR spectra relative to water. In other words, fat protons spin slightly faster than water protons. Thus, when a gradient echo image is formed, the signal coming from the fat accrues slightly more phase than the water does over the same time period. This causes the fat to appear slightly shifted in position from the water along the direction of the readout gradient. The phenomenon is partially resolved in spin echo images because the inversion pulse causes the fat's phase to be inverted, so fat's positive phase offset becomes a negative offset and it "catches up" to the water when the echoes form.

Fold-Over Artifacts

If you look at an MR image and notice that it seems like data from one side of the image are appearing on the other side, you've found a fold-over (or aliasing) artifact. These artifacts occur when the field of view is smaller than what is being imaged. The frequency or phase of something outside the field of view will be incorrectly encoded. Because Fourier transforms assume that the data are periodic, a frequency peak that is outside the expected frequency range is mathematically identical to a different (and incorrect) frequency peak in the range, leading to the placement of that item on the other side of the image. These artifacts can be resolved using a larger field of view or specially designed "spatial saturation" pulses that can nullify the signal coming from outside the region of interest.

Flow Artifacts

When fluid is flowing while the RF pulses are being played, the fluid will be excited in one location and then continue moving prior to the formation of an echo. As a result, the signal will appear to come from a different location when the echoes form. In a 2D gradient echo sequence, through-plane flow will usually appear as multiple vessels replicated along the phase encoding direction. In-plane flow will often appear as a shift of a blood vessel. A variety of techniques to compensate for flow artifacts can be applied to various situations.

Intensity Variations

There are cases in which two tissues in an image that should have an identical appearance will have different brightness. This is frequently caused by B_1 inhomogeneity. The root cause is that the length of the RF pulse is similar to the size of the MRI bore, making it difficult to design a pulse to excite tissue evenly. This leads to a spatial variation in the excitation, so tissue in one area has a slightly lower flip angle compared to another area, causing an intensity variation. Furthermore, the coils that receive the data may not sense each area equally. Taken together, the resulting image may show a spatial intensity variation. It is a larger problem with 3T scanners, especially with brain images.

Extrinsic Artifacts and Metal Artifacts

Sometimes, the artifacts do not come from the pulse sequence or the tissue but, rather, from the scanning environment. Many of these cases relate to metal. Some metal artifacts, such as those from large metal implants, cause the signal in the neighboring tissue to be completely obliterated, leaving a big black hole in the image. It is also common to see shape or contrast disturbances in images due to patient clothing. Bra underwires and metal zippers are obvious sources of error, but things such as glitter and metallic thread are more difficult to notice but can lead to image fuzziness or nearby distortions. Some makeup and hair products use glitter, which can cause subtle (or not so subtle) changes, especially in brain images.

Even farther from the scanner, artifacts can be introduced. For example, if the scan room door is not closed or a device such as a cell phone is left in the scanning room during the scan, the RF emitted can cause spikes at particular frequencies in k-space, leading to lines in the image known as zipper artifacts.

SAFETY

MRI is a very safe imaging modality. Because it does not use ionizing radiation, there is no risk of DNA damage leading to neoplasms. Nonetheless, there are some important safety precautions that must be taken to avoid injury due to ferromagnetic objects as well as heating.

Ferromagnetism, or Things Flying Across the Room

The most obvious risk with MRI is that a piece of metal that is ferromagnetic is brought into the scan environment. If you enter the scanner with a belt buckle that is magnetic, you will feel the odd sensation of the buckle being pulled away from your body by the magnetic field. On the other hand, if you accidentally wheel a steel oxygen cylinder into the room, it will fly across the room into the bore of the magnet, which can cause serious injury and death by colliding with people in its path or pinning a patient in the scanner. The magnet is powerful enough to pick up a hospital bed or wheelchair and suck it into the bore. Smaller pieces of ferromagnetic material may not injure anyone but can cause field variations that degrade the images. Therefore, it is important that no ferromagnetic objects are introduced into the scanning environment.

Implants

In addition to exerting force on large objects, the scanner can have effects on smaller, implanted devices. For example, some neural clips and other implants are ferromagnetic. There have been cases of patients with ferromagnetic aneurism clips entering the scanner and subsequently suffering brain damage or death because the magnet exerts a force on implant. Furthermore, the RF energy used to excite tissue can also cause heating. If a conductive object such as pacemaker lead is the right length, it will act as an antenna for the RF signal, concentrating the energy in a very small amount of tissue. This can lead to severe burns. Thus, it is important to screen for any implanted devices and check the manufacturer's specification for an "MR Conditional" specification that explains what conditions are safe for that device to be scanned, including field strength, RF, and gradient limits.

For active implants, such as pacemakers and nerve stimulators, many of the modern devices are approved for MRI. However, they vary in their scan parameters, and some devices must be disabled or interrogated after the scan to ensure proper function.

Indwelling Metal

Patients will often present for scans with indwelling metal that is not a medical implant. This includes piercings, bullets, and small metal particles in the eye as is common with machinists. Some tattoo inks are made with iron oxide or other metals, making them conductive. Often, this is not an issue, but it is important to assess each subject's case before placing them in the MRI. In cases in which these items cannot be removed, special care must be taken to prevent burns or tissue damage from ferromagnetism.

Concerns Regarding Contrast

It is common for patients to be worried about the safety of MR contrast agents, most of which are gadolinium-based. These concerns often range from worries about deposition in the body to the effects on kidneys. Although the contrast agents available today are generally considered to be safe, many institutions will withhold contrast in patients with reduced renal function.

Peripheral Nerve Stimulation

When the gradient waves are particularly strong, it is common for patients to experience peripheral nerve stimulation, a phenomenon in which peripheral nerves are activated directly by the scanner. It is commonly described as a tingling, frequently in the small of the back. It is not dangerous but can be disconcerting. It can be remedied by reducing the rate of change of the gradient waves if the pulse sequence being used will allow it.

CLINICAL CORRELATIONS

In clinical imaging, there are some rules of thumb worth noting. Most of them make logical sense given that clinical MRI images hydrogen, primarily water followed by fat. The rules are as follows:

- Water is bright on T_2 and dark on T_1. Because water has a long T_1, scans with a short TR lead to a relatively short longitudinal magnetization vector for water because it does not have time to recover before subsequent excitations and it appears dark on

T_1-weighted scans as a result. On the other hand, water has a long T_2 value, so its transverse magnetization remains strong even with longer TEs, causing it to appear bright on T_2-weighted images.

- Inflammation leads to increased fluid, which leads to areas of hyperintensity on T_2 images and hypointensity on T_1 images. This makes perfect sense following the logic of the prior bullet point.
- White and Gray matter appear with different signal intensities in various sequences, reflecting their core composition differences. Most predominantly, myelinated white matter has significantly higher lipid content and less cytoplasm (fluid).

In T1 sequences, fluid (CSF) is dark and white matter is brighter (hyperintense) than grey matter.

In T2 sequences, fluid is bright and grey matter is brighter than grey matter

In FLAIR (Fluid Attenuation Inversion Recovery) sequences, fluids are "nulled" so fluid is dark and grey matter is brighter than grey matter.

Figure 11.4 shows several sequences of an MRI from a normal patient.

Multiple Sclerosis

Multiple sclerosis (MS) is an inflammatory disease process of the brain that leads to the destruction of myelin in the central nervous system. Figure 11.5 shows MRI sequences of a patient with MS. The disease demonstrates some characteristic MRI findings that flow from the disease process which emphasize the effects that inflammation has on T_1, T_2, FLAIR, and diffusion-weighted images. These effects can be mostly conceptualized by considering that with inflammation, there will be an increase of free fluid in the diseased regions. On T_1-weighted images, the MS lesions will appear to be hypointense. This makes sense because the increased fluid in the lesion, with its long T_1, will appear darker on T_1 scans than areas that

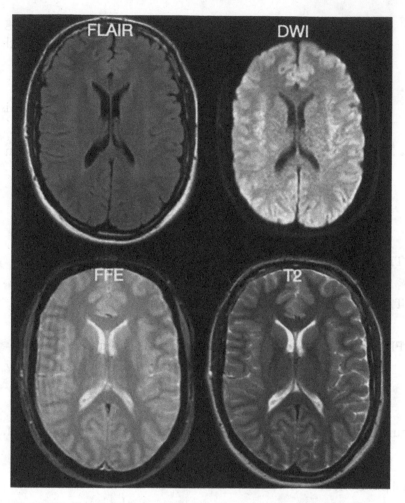

Figure 11.4 MRI of a normal patient. (*Top left*) FLAIR, which searches for a broad spectrum of abnormalities, including edema and acute or chronic damage. (*Top right*) DWI, which searches especially for acute stroke. (*Bottom left*) Fast field echo (FFE), which searches for bleeding. (*Bottom right*) T_2, which searches for a broad range of structural brain lesions.

do not have increased fluid. As is also expected, the T_2 images will show hyperintensities in the same region due to the fluid's longer T_2 value. This is further accentuated with T_2–FLAIR imaging, which nullifies the CSF in the ventricles while still showing the hyperintense lesions.

Due to the increase of unrestricted fluid motion from the inflammation and destruction of the myelin, readers may expect that MS would

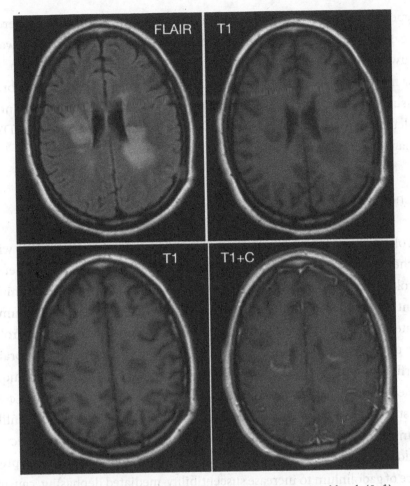

Figure 11.5 Multiple sclerosis. Top frames are from the same axial level. (*Left*) FLAIR, which shows lesions well, mainly chronic. (*Right*) T_1 image, which does not show lesions well. Bottom frames are from the same axial level. (*Left*) T_1 without contrast and (*right*) T_1 with contrast, which highlights the areas of active demyelination.

lead to hypointensities on DWI images—after all, the increased apparent diffusion should cause more signal loss when the diffusion gradients are played. For that matter, MS does demonstrate increased ADC, as would be expected of the increased unrestricted fluid, something that seems like it should be inversely related to signal intensity on the DWI images. Somewhat counterintuitively, the DWI scans will demonstrate areas of hyperintensity. This is known as the "T_2 shine-through" effect, whereby the increase in T_2 of the signal from the increased presence of water outweighs the losses in the DWI images due to increased diffusion. So, for many b-values, the MS lesions will appear brighter than healthy brain tissue despite the increased diffusion. For this reason, it is important to calculate the ADC rather than assuming that a hyperintensity on a DWI image necessarily means decreased diffusion.

Ischemic Strokes

Strokes are another commonly investigated occurrence with MRI, with ischemic strokes representing approximately 80% of the total. Ischemic strokes are commonly investigated both with and without gadolinium contrast and demonstrate how certain MRI sequences can change immediately, whereas others evolve more slowly. In an acute ischemic stroke, the underlying cause is an occlusion to an artery feeding part of the brain. Perfusion in the brain can be estimated by using a T_2^*-weighted technique called *dynamic susceptibility contrast*. In this technique, gadolinium contrast is injected and a fast T_2 sequence is repeated multiple times. Unlike many contrast-enhanced sequences that use the T_1 shortening effect of gadolinium to make blood appear brighter, these sequences use the presence of gadolinium to increase susceptibility-mediated dephasing, causing a loss of signal. When the perfusion maps are calculated, the areas of the brain downstream of the arterial blockage will immediately demonstrate reduced perfusion. DWI shows restricted diffusion when the stroke is relatively acute, at a time when other sequences may miss the ischemia. This is illustrated in Figure 11.6.

However, not all changes happen immediately, a fact that allows for estimates of the start time of the stroke to be made. FLAIR images taken following a stroke will eventually show increased signal in the region of the reduced perfusion due to vasogenic edema; however, this process takes 3–6 hours in most cases, with some reports of negative findings on FLAIR images for up to 24 hours following a stroke.

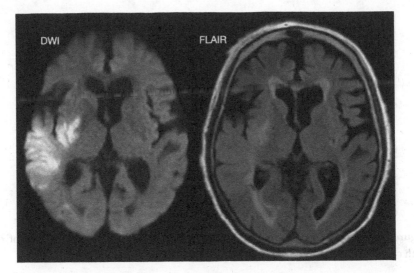

Figure 11.6 MRI with acute stroke. (*Left*) DWI, which shows restricted diffusion in a portion of the MCA distribution. (*Right*) FLAIR, which shows only subtle signal change and fullness of tissue in the stroke area.

Hemorrhagic Stroke

Most strokes are ischemic, but approximately 13% are hemorrhagic. They can be primary hemorrhage, such as from hypertension, or secondary to blood vessel damage from ischemic stroke. This is especially a concern after an infarcted area has reperfused due to either natural clot lysis or reperfusion therapy with thrombolytic agent or endovascular procedure. Chapter 2–6 showed that blood is radiodense on CT. On MRI, the appearance depends on the sequence. We use fast field echo mainly to identify hemorrhage from other pathology. Figure 11.7 shows hemorrhage with comparison of the MRI and CT findings.

MRI is the imaging modality of choice to assess for cerebral amyloid angiopathy (CAA) given its much higher sensitivity for microhemorrhages, which can often go unseen on CT. This is not due to spatial resolution but, rather, MRI's more unique ability to bring out variations in tissue properties. T_2^*-weighted imaging sequences such as Gradient Echo (GRE) or Susceptibility weighted imaging (SWI) on MRI intentionally take advantage of an inherent "artifact" to identify new and old areas of

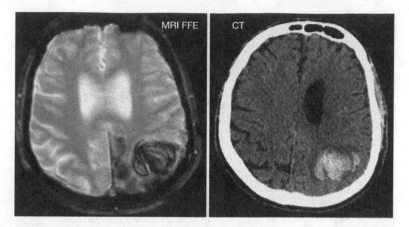

Figure 11.7 Hemorrhage seen on MRI and CT. Left occipital hemorrhage in a patient with cerebral amyloid angiopathy. FFE, fast field echo.

intracranial bleeds. The deposition of iron-rich hemoglobin results in magnetic field variances and ultimately a susceptibility blooming artifact to which T_2^* sequences are particularly sensitive. This phenomenon is also seen in other mineral deposition such as calcium. The image will display small black dots that are signal voids but very useful for the provider to identify multiple microhemorrhages that may have been clinically silent. As opposed to hypertensive microhemorrhages, CAA should typically spare the basal ganglia and pons.

Larger intraparenchymal bleeds can be apparent on T_2 sequences, lesser so on T_1. Just as the calcium-rich cortical bone of the skull is not visualized (only marrow), hemosiderin is black (hypointense). The surrounding edema, however, will have a slightly more intense signal than surrounding brain tissue on T_2. The appearance on T_1 and T_2 sequences will vary in accordance with the chronicity of the bleed: hyperacute, acute, subacute, and chronic. This is due to the changing composition of the hemorrhagic bed as the blood cells are broken down.

For comparison, CT is a rapid tool for identifying intracranial hemorrhages. In CAA, the amyloid plaque deposition leads to friable blood vessels throughout the cortex, leaving them vulnerable to rupture. The majority of the time, this may be a clinically silent microhemorrhage.

A common scenario in which CAA is diagnosed is when a patient presents for an acute headache, with or without neurological deficits, and is found to have an intraparenchymal hemorrhage in an area of the brain atypical for hypertensive bleeding. CT is based on radiopacity, and the accumulation of iron-rich hemoglobin attenuates more X-rays than the surrounding brain tissue, giving it a higher average Hounsfield unit (HU) value and appearing brighter, although not nearly as bright as the cortical bone of the skull. Associated edema, on the other hand, has less protein than soft tissue and will have an HU value closer to that of water (0) and therefore be darker by comparison, although not as dark as the CSF within the ventricles.

Nuclear Imaging

EVAN M. JOHNSON ■

WHAT IS POSITRON EMISSION TOMOGRAPHY/SINGLE-PHOTON EMISSION COMPUTED TOMOGRAPHY?

Positron emission tomography (PET) and single-photon emission computed tomography (SPECT) are two forms of radio-isotope–based imaging modality. Compounds that include a radioactive element are introduced to the body, and images are produced from the detection of emitted photons. It is common for a second imaging modality, usually CT, to be obtained and overlaid to provide anatomical mapping to the nuclear image.

How Did PET/SPECT Come About?

Scientists have worked to harness radiation since the discovery of its existence, with the potential for medical imaging being among the first applications realized.

As the late 1800s gave way to the 20th century, French physicist Henri Becquerel discovered that uranium naturally emitted rays that could penetrate paper, be deflected by magnetic fields, and deliver energy when absorbed. Marie Curie would later coin the term "radioactivity," share a Nobel Prize with both her husband Piere and Henri Becquerel for the discovery of this, and would pioneer the identification of elements that naturally produce radiation such as radium. In 1911, she became the first recipient of two Nobel Prizes.

The use of photographic plates opposite a radiation source to help identify tumors and other lesions in the early 20th century would herald the advent of medical imaging. This continues to be the basis of X-ray and CT imaging today. The evolution from externally sourced to internally sourced radiation would soon follow.

In 1913, Frederick Proescher published studies on the intravenous injection of radium into patients with select diseases such as pernicious anemia and arthritis. He was interested in potential therapeutic effects but helped establish an understanding of tolerable and lethal levels.

Hungarian George de Hevesy received a Nobel Prize in 1944 for the development of radiotracers—coupling a non-radioactive compound with a radioisotope for study purposes. This allowed him to show plant uptake of materials, then in animal studies, and later, using radioactive phosphorus, he demonstrated dynamic remodeling of bone in humans. The ability to study biological function through radiotracers was the foundation of modern nuclear imaging. The first papers describing this use in diagnostic imaging were published in 1939.

Following this, advances were made in the production of particular radioisotopes and, later, techniques and technology for the accurate detection of the emitted radiation. A key part of this was the development of scintillation detectors. Simply stated, scintillators absorb radiation and convert that energy into light. They were first created at the turn of the 20th century, but the subsequent development of and coupling with photomultiplier tubes led to modern scintillation detectors that could be the backbone of nuclear imaging systems. Benedict Cassen, Lawrence Curtis, Clifton Reed, Raymond Libby, Gordon Brownell, H. H. Sweet, Hal Anger, and David Kuhl were among many in the 1950s who pushed this technology and paved the way for diagnostic nuclear imaging systems. In fact, the work of Kuhl to develop tomographic imaging is thought to have informed the development of X-ray–based CT almost a decade later.

During the next two decades, commercially developed radioisotopes and detectors continued to be refined along with their medical applications.

In 1973, David Goldenberg provided a breakthrough by demonstrating the use of radiolabeled carcinoembryonic antigen antibodies to detect cancerous tissue in the body. His group coupled iodine-125 and I-131 to goat IgG antibodies against carcinoembryonic antigen and injected them into Syrian hamsters that had received tissue grafts of colon cancer. They were then able to demonstrate the selective uptake of the antibodies using a scintillation counter. Goldenberg translated this work into human demonstrations by 1978. The use of antibodies to target specific tissues would come to be employed for both diagnostic and therapeutic purposes, as radioisotopes with sufficient activity could lead to local cell death.

John Keyes developed the first SPECT camera in 1973, followed by the development of a dedicated head SPECT camera by Ronald Jaszczak in 1976. Their work introduced a system featuring a camera that would rotate around the patient. By this time, computers were able to facilitate the creation of axial image slices from the 360-degree raw data (filtered back-projection).

During the 1980s and 1990s, as SPECT applications in clinical medicine evolved, positron emission systems emerged. Coincidence detection took advantage of the positron generation of two photons propelled 180 degrees from one another. Ring detectors could then capture the near-simultaneous photon collisions and use these data to help partially localize the point of origin. By the turn of the millennium, PET scanners were validated as a cancer screening tool for both detection and verification of successful treatment.

What Sets PET/SPECT Apart from Other Imaging Modalities?

Whereas most other medical imaging modalities are inherently "anatomical" or structural scanners, nuclear imaging is more tailored for assessing functional information or tissue-specific localization. Depending on the desired application, radioisotopes can be attached to tracers that are either taken up as a marker of cellular metabolism or antigens designed to affix antigens associated with cells of interest. SPECT and PET typically require an additional anatomical scan (CT or magnetic resonance) for overlay.

Radioisotopes used for diagnostic purposes may also be used to deliver therapeutic localized radiation. Whereas focused ultrasound can be applied for targeted ablation procedures, CT and magnetic resonance imaging (MRI) are strictly diagnostic modalities. X-ray–based radiation therapy also exists, although this is a carefully planned procedure that deviates significantly from traditional X-ray radiography.

HOW DOES PET WORK?

Positron emission tomography is carried out by introducing a positron-emitting radioisotope to the body via a compound designed to allow a relatively safe targeted tissue uptake with slow clearance to allow continuous generation of photons from the emitted positrons that can be detected and form the basis of three-dimensional images. An even simpler way of thinking about it is as a Trojan horse compound introduced to plant a homing beacon that permits assessment for possible abnormal tissue of interest.

A positron is an antiparticle opposite of an electron, possessing the same mass but a positive charge rather than negative. It is generally a short-lived decay product formed when a proton becomes a neuron, known as beta decay (specifically, β^+ decay). Both alpha and beta decay are illustrated in Figure 12.1. Beta decay can be seen naturally in some elements such as potassium-40. Commercially, fluorine-18 is created by bombarding oxygen-18 with a high-energy proton beam generated in a cyclotron. In this reaction, a high-energy proton can eject a neutron and take its place, creating F-18. O-18 is a naturally occurring isotope, although it makes up only approximately 0.2% of the oxygen in the atmosphere or water. Techniques exist to produce "enriched" O-18 water, which is the common substrate for the creation of F-18.

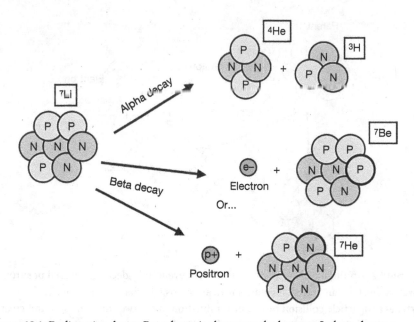

Figure 12.1 Radioactive decay. Beta decay is shown on the bottom. In beta decay, a radioisotope "parent" decays from a higher to lower energy state "daughter." In β⁻ decay, the parent atom, a neutron within the nucleus, becomes a proton, releasing an electron and antineutrino particle. In β⁺ decay, the parent atom, a proton within the nucleus, becomes a neutron, releasing a positron and neutrino particle. Alpha decay is shown for completeness but does not apply to this context.

The radioisotope then decays, producing the short-lived positron. Positrons will be released omnidirectionally. They travel only a very short distance, typically less than 3 mm, before being attracted to an oppositely charged electron and colliding with it. This is an annihilation event and produces mirror gamma ray photons of equal energy that travel 180 degrees apart from one another. This is illustrated in Figure 12.2.

A PET scanner utilizes a fixed ring of scintillation detectors to capture photons over a period of time. The near-simultaneous arrival of coupled gamma rays at opposing scintillators is called a coincidence detection, and the accumulation of these data is used in back-projection to display the distribution of radioisotope throughout the body.

A CT (or MRI) scan is commonly used to overlay anatomy over the PET scan.

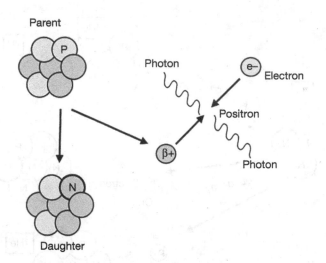

Figure 12.2 Annihilation event. As a subsequent event to β⁺ decay, a released positron travels a short distance until it collides with an oppositely charged electron. This particle–antiparticle collision results in annihilation of the two initial masses and creates two gamma photons of equal energy traveling 180 degrees apart.

What Are the Components of a PET System?

A modern PET scanner consists of a large ring array detector, made up of scintillator crystals closely coupled with photosensors, as shown in Figure 12.3. The detection and conversion of gamma rays to light and digital quantification mirrors that of X-rays in a CT scanner. The key difference is

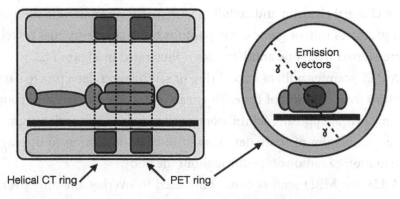

Figure 12.3 PET Scanner. Modern PET systems combine a detector ring, composed of scintillating detector elements, with a helical CT gantry, allowing for both scans to take place in a session to optimize overlay positioning.

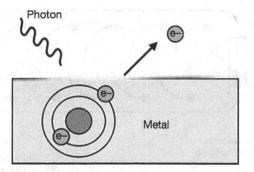

Figure 12.4 Photoelectric effect. The photoelectric effect describes the phenomenon in which a high-energy photon interacts with an object and a collision with an orbiting electron imparts sufficient kinetic energy to release the electron from orbit.

that modern PET scanners can take advantage of the patient's internalized source of radiation and use a fixed 360-degree ring array of detectors rather than requiring a rapidly rotating gantry.

A gamma ray reaches the scintillator crystal, which creates a photon of light. This light then strikes a transmission photocathode, usually a specialized alkali metal that converts light photons into free electrons via photoelectric effect that it ejects as a beam within a photomultiplier tube vacuum toward a nearby anode. The photoelectric effect, in simplest terms, is when sufficient energy from a photon is absorbed by an electron so that it has greater kinetic energy than its binding energy keeping it in orbit within an atom, making it a free electron. This is illustrated in Figure 12.4.

A photomultiplier tube affixed to the back of the scintillator thus receives the photon and takes its energy to elicit an electron beam current through a maze of dynodes across the tube, as shown in Figure 12.5.

A dynode is an electrode made of a metal plate with a specialized surface that easily emits electrons when excited, as in this case by a collision with another electron. The photon begins at a photocathode, and the excited electrons are pulled toward the anode at the far end, where the photomultiplier sends the generated current through cables to the computer, typically first a dedicated coincidence processing unit. Because photons are generated as 180-degree coupled emissions, the PET scanner is able to detect the near-simultaneous arrival of these particles, and software is able

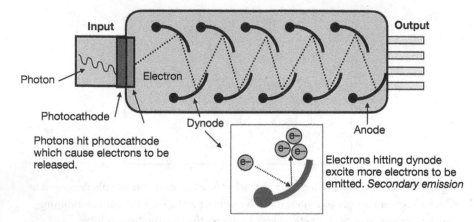

Figure 12.5 Photomultiplier tube. A photomultiplier tube (PMT) is a vacuum tube that accepts an initial light photon or photon beam which it converts to an electrical signal. The light photon first interacts with a photocathode, where electrons are released via a photoelectric effect and immediately repelled by the photocathode charge and drawn to the anode placed at the opposite end of the photomultiplier tube. A series of dynodes are placed throughout the tube to ensure repeated electron collisions. Each collision releases additional electrons from the dynodes, which are a metal with loosely held electrons. These collisions act to "nsure re the electron signal before it reaches the PMT anode and transmitted via connector pins to wiring to the central processing unit workstation.

to log these two specific spatial points as an assumed pair. This allows the software to assume that the origin of the particles was located along the line between the two detector elements. This is considered coincidence detection. Over the course of the period of time in which the detector is active, line after line of detected photon pairs helps triangulate the areas of the body with the greatest accumulation of radioisotope.

It is common now for a dedicated CT scanner to be combined with the far end of a PET system to allow anatomical overlay. Some systems have a coupled MRI, which requires the patient bed to cross into a shielded room.

What Are the Basic Tracers and Applications of PET?

Like SPECT, the imagination and capability of chemists and biochemists are the limit of the possible tracers that can carry a positron emitter.

By far the most well-known PET compound is fluorodeoxyglucose (FDG). Glucose needs no introduction, nor does its central role in human metabolism. Deoxyglucose is an interesting compound, in which a hydroxyl group is replaced by a proton. This minor change prevents complete breakdown of the molecule in glycolysis. Cells can readily uptake this compound, but the inability to metabolize it into water and carbon dioxide arrests it in the cell, making it an ideal Trojan horse for medical imaging. FDG is a radiotracer analog of this, with an F-18 radioisotope in place of a proton group. The F-18 must be created via a cyclotron.

In practice, injection of FDG results in swift uptake of the compound by tissues with high glucose demand. The brain, heart, liver, and spleen are all such organs at baseline. The renal system biologically clears FDG, so it can also accumulate in the bladder. Primarily, FDG-based PET imaging is used as a cancer screen. The vast majority of tumor cells are rapidly dividing and aggressively transport nutrients, including glucose. So although FDG uptake verified in PET imaging is not a particularly specific finding, it is a very sensitive one for malignancy.

FDG can also help reveal regional areas with poor metabolism. One such application is diagnostic imaging of dementia. Certain conditions are associated with hypometabolism in different areas of the brain. FDG PET imaging in Alzheimer disease commonly shows reduced uptake in the temporal and parietal lobes of the brain, in the frontal and temporal lobes in frontotemporal dementia, and in the occipital lobes in Lewy body dementia.

In cases of refractory epilepsy in which there is a suspected local focus from which seizures originate, FDG PET can be used to help localize this focus and facilitate tissue ablation. If the patient is carefully monitored in a medical facility and a seizure occurs, the patient can be immediately taken for an FDG PET scan. This timing is paramount because a post-ictal patient should have reduced metabolic activity in the cerebral seizure focus. FDG PET can also be used to confirm medial temporal sclerosis as an underlying cause of epilepsy, demonstrating asymmetric uptake and not requiring special timing.

Although FDG PET is common, given the relative simplicity of design and broad range of applications, numerous other radioisotopes and tracers are available for more specific interests. Other positron emitters used in PET studies include the following:

- Carbon-11
- Nitrogen-13
- Oxygen-15
- Manganese-52
- Cobalt-55
- Copper-64
- Gallium-68
- Rubidium-82
- Zirconium-89

Many of these isotopes have half-lives considerably shorter than F-18's 110 minutes, putting considerable time-sensitive pressure on the production and utilization. The half-life of Rb-82 is less than 90 seconds! On the other end of the spectrum, Zr-89's half-life is greater than 3 days, and that of Mn-52 is nearly 6 days. This is advantageous in certain respects.

Biochemists take advantage of multiple aspects of Zr-89, created from proton bombardment of yttrium-89, for the radiolabeling of antibodies. The chemical characteristics of zirconium make it an ideal candidate for chelation and affixation to antibodies selected for specific tissues of clinical interest. The longer half-life of 3.25 days is ideal for this application because circulating antibodies commonly require 2–4 days for optimal pharmacokinetics when targeting cancerous cells.

O-15 is commonly used to measure blood flow, particularly to areas such as the brain, heart, and tumors.

Cobalt-55 is similar in use to FDG because it is a Trojan horse mimic for calcium. It can be used to visualize areas of high calcium influx, such as damaged tissue in a stroke bed or necrotic tissue within a tumor.

In all cases, medical physicists help oversee the process and the careful calculations required. The radioactivity or the employed isotope has to be carefully accounted for and involves an understanding of not only the amount injected into the patient but also the timing. After a set predetermined amount of time is given for uptake, the PET scanner begins to acquire data over another predetermined amount of time specific to the tracer, isotope, and application.

HOW DOES SPECT WORK?

What Are the Components of a SPECT System?

Single-photon emission computed tomography is a similar modality in that a radioisotope chemically affixed to a biomolecule is injected into the body, taken up in a tissue of interest where radiation is emitted and captured by a surrounding detector array. This is then used to generate a cross-sectional image that can be overlaid upon an anatomical CT scan to show the particular areas of the body the tracer accumulated.

The key difference between SPECT and PET is that SPECT utilizes different radio-isotopes (and ligand tracers). Single-photon emitters such as I-123, technetium-99m, xenon-133, thallium-201, and F-18 are examples.

I-123 decays to tellurium-123 with the emission of a 159-keV gamma ray and a half-life of approximately 13 hours.

99mTc has an "m" designation because it is a metastable nuclear isomer of Tc-99 and is the most used radioisotope in global medical imaging. It decays to Tc-99 with a 6-hour half-life. 99mTc is itself a decay product of molybdenum-99 (almost 3-day half-life), which permits off-site generation and shipping.

Xe-133 is a gas and most often inhaled to facilitate imaging of the lungs, decaying to caesium-133 with a half-life of approximately 5 days.

Thallium-201 has a 73-hour half-life and is most often used for cardiac imaging. This radioisotope itself can mimic potassium in the human body, which facilitates its uptake in tissues of interest where it emits 68–80 keV photons.

Not exclusive to PET, F-18 can also be used in SPECT scanners, commonly as FDG for similar applications.

What Are the Basic Tracers and Applications of SPECT?

Technetium-99 is a very common radioisotope for SPECT imaging and can be utilized for several specific targets. Tc-99m methylene diphosphonate is a radio-coupled compound that has a high affinity for sites of new bone formation. This compound is a key radiotracer for SPECT-based bone scans, which can demonstrate areas of abnormal or pathological bone turnover.

In myoperfusion studies, 99mTc can be coupled with tracers such as tetrofosmin or sestamibi. After injection of this radiolabeled tracer, the patient may undergo a cardiac stress test, either by participating in a demanding exercise, commonly using a treadmill, or by administration of a medication that induces increased heart rate. SPECT imaging in this case is used to identify areas of myocardium that do not take up expected levels of tracer, which indicates poor blood flow and is suggestive of ischemia.

Although most commonly used in cardiac stress tests, 99mTc-sestamibi is also used to evaluate the parathyroid gland. Sesta derives from the Latin *sexta*, or six, and sestamibi is appropriately named for its six methoxyisobutylisonitrile (MIBI) groups that are attached to the 99mTc radioisotope. Because that can be a mouthful, it is typical to refer to SPECT imaging using this radiotracer as a MIBI scan. Sestamibi is a lipophilic compound that diffuses into cells with a high number of mitochondria, such as myocardium. This property means that the tracer is additionally optimal for uptake into mitochondrial-rich parathyroid tissue. A MIBI scan of the parathyroid can reveal underlying lesions or tumors.

Iobenguane, or MIBG, is a different tracer typically used for imaging of adrenergic tissue. It is usually coupled with I-123 or I-131. This compound accumulates in adrenal medullary chromaffin cells as well as in presynaptic adrenergic neurons, enabling its use in SPECT-based screening identification of certain endocrine tumors, especially those of the adrenal glands including neuroblastomas. This is a good example of a radiotracer that can serve dual diagnostic and therapeutic roles. MIBG can be coupled with I-131 to provide radio-ablation to cancerous tissue while simultaneously allowing gamma imaging. I 123 decay to Tellurium-123 emits a gamma ray with a kinetic energy approximating 159 keV. I-131 decay, however, is an 8-day half-life beta decay that produces a high-energy electron, antineutrino, and gamma emissions to become xenon-131. The production of electrons with kinetic energy ranging from 250 to 800 keV is the aspect of this decay that imparts the most damage to surrounding tissue.

CLINICAL CORRELATE—PARKINSON DISEASE

In Parkinson disease (PD), there is an insidious slowly progressive intracellular accumulation of alpha-synuclein Lewy bodies within certain neuronal cells, impairing function. Among the affected cells, the substantia nigra experiences a gradual degradation of dopaminergic cells that are critical to the proper functioning of the basal ganglia. This brings about the cardinal motor features of PD: asymmetric resting tremor, slowness (bradykinesia), limb (ratcheting) rigidity, and diminished lift and stride while walking.

The initial patient evaluation in clinic may not always be resoundingly obvious because the patient may not have all features early in the course, may have atypical presentation, or may have one of several parkinsonian mimic conditions (parkinsonism). Although PD is a clinical diagnosis, the use of SPECT imaging may help the provider be more certain in cases in which the ultimate diagnosis is not clear.

In Chapter 4, we discussed termination of synaptic transmission, and one mode is reuptake of the transmitter back into the nerve. This is the case with dopamine. It binds to the dopamine transporter protein, which is used to pump dopamine out of the synaptic cleft and back into the cytoplasm, where other transporters repackage the dopamine for reuse.

A dopamine transporter SPECT study (DaTscan) assesses the activity level of pre-synaptic dopamine transporters in the brain via injection of ioflupane, a nonfunctional cocaine analog coupled to an I-123 radio-isotope. Typically, there is robust uptake in the striatum, resulting in a comma-shaped signal bilaterally. Figure 12.6 shows a normal DaTscan with a high amount of signal in the striatum. In idiopathic PD, there would be an expected asymmetric reduction of overall signal and transition from a comma shape to a dot or oval representing the caudate nucleus.

Note that this scan is not specific for PD because it does not discern between the various Lewy body disorders, which also include multiple system atrophy and dementia with Lewy bodies. Progressive supranuclear palsy is a "Parkinson-plus" tau disorder, rather than alpha-synuclein, but also has a positive DaTscan with reduced uptake, albeit with subtle differences from PD that may or may not be apparent. Numerous medications can cause a false-positive DaTscan and should be carefully

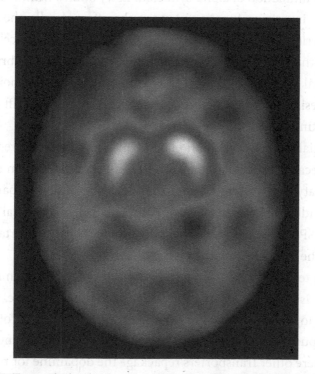

Figure 12.6 DaTscan. The bright signal on either side of the center is the uptake of the radiotracer ioflupane in the striatum in a patient, highlighting the dopamine transporter protein of intact neurons.

screened for in clinic before ordering the study. Most of these are psycho-tropic medications, including stimulants and antidepressants, and need to be held 1–3 days in advance, depending on the pharmacokinetics.

ARTIFACTS

Most artifacts in nuclear imaging are related to their anatomical scan counterparts. Artifacts in the corresponding CT scan, such as respira-tory motion or metal-associated artifacts such as streaking, have been described previously in this book.

An inherent obstacle to high-resolution localization in PET is that photons are not directly created within the radioisotope. The isotope emits positrons, which then travel a limited distance before annihilation creates the paired photons. This creates an inherent "blurring" to the localization of the accumulated tracer, although normally not so much as to lead to an indeterminate result.

However, PET is established to have superior resolution than SPECT for additional reasons. Coincidence detection not only improves the ability to triangulate the source of radioactive decay but also facilitates exclusion of noise. Scatter radiation, or simply any single photons that do not di-rectly represent gamma rays of the radioisotope decay, can be filtered out in PET imaging because they would not have a near-simultaneous paired photon detection. This cannot be applied to SPECT. SPECT imaging gen-erally requires longer time of acquisition and increased patient radiation burden.

Although not an artifact per se, certain considerations must be taken into account with nuclear imaging. A key variable in both SPECT and PET imaging is carefully monitoring and standardizing the radioactivity of the radiotracer given to the patient. Medical physicists usually oversee this process, which requires careful calculations and timing, beginning with the generation of the radioisotope. Depending on the timing of ad-ministration, the amount of isotope administered, and the time of data ac-quisition, the generated image may inadvertently and inaccurately suggest increased or decreased uptake. This is akin to under- or overexposure of camera film.

An additional concern for FDG PET imaging is any alterations to the patient's glucose metabolism. Patients are asked to fast in the hours preceding the scan and abstain from medications such as insulin that could alter uptake of FDG.

KEY EQUATIONS

Activity (decay rate) of an isotope:

$$A = \lambda N$$

where A is activity, given in Becquerel (Bq), or s^{-1}; λ is the decay rate constant (specific to the isotope), also in Bq; and N is the number of atoms remaining.

Half-life of a radioisotope ($T\frac{1}{2}$), the time it takes half the atoms to decay:

$$T\frac{1}{2} = 0.693/\lambda$$

Radiation flux (I):

$$I = I_0 e - \mu x$$

where I is the initial number of atoms, I_0 is initial flux of radiation; μ is the linear absorption coefficient, and x is the thickness of the substance.

Photoelectric effect:

K_{max} = (Plank's constant × frequency of the electromagnetic wave) –
the minimum energy required to free an electron from orbit

where K_{max} is the maximum kinetic energy of the ejected electrons observed.

Laboratory Studies

KARL E. MISULIS ■

Laboratory studies have been key to neurologic diagnosis for decades, and they have gotten more complex over time. Similarly, their interpretation has become increasingly complex. Sometimes it takes several studies to make a diagnosis, or a study may be only one clue to making a diagnosis. This chapter presents a discussion of some of the key types of laboratory studies, mainly body fluid and pathology because neuroimaging and neurodiagnostics are discussed extensively in separate chapters of this book.

ROUTINE LABORATORY STUDIES

Routine analytic laboratory studies are important mainly for their identification of medical conditions that have neurologic implications. For many tests, there are multiple methods of analysis, but generally one is discussed for illustrative purposes.

Glucose dyscontrol is a common cause of encephalopathy and even focal signs. There are several methods for analysis, and these differ between point-of-care devices and central labs. Most methods use glucose oxidase, which is an enzyme that catalyzes the metabolism of glucose into D-glucono-1,5-lactone + hydrogen peroxide. The hydrogen peroxide in the presence of horseradish peroxidase reacts with a chromogen, which then creates a color that can be measured. Another enzyme that can be used is hexokinase, which converts glucose plus adenosine triphosphate into glucose-6-phosphate and adenosine diphosphate. The glucose-6-phosphate in the presence of nicotinamide adenine dinucleotide (NAD) plus a dehydrogenase enzyme produces NADH, which can then be measured using a spectrometer. The glucose oxidase and hexokinase methods can give different results with the same patient, so it is important to be consistent when measuring glucose.

Ion levels (sodium, potassium, calcium, chloride, etc.) are often measured using ion-specific electrodes. These allow detection of one or a small selection of ions by measuring the potential difference between an ion-specific electrode and a reference electrode in equilibrium conditions. A specific and narrow range of values can be measured, and the results can be calibrated with known ion levels.

Hematology analysis is often performed by differential cell counters. There are different types, but one type sends blood cells one at a time through an opening to a zone where they are illuminated by laser light. The characteristics of the light scatter, and absorbance correlates with blood cell type.

Prothrombin time (PT) is a measure of blood clotting on the extrinsic pathway (usually triggered by external trauma). This is measured on blood collected in a citrate tube to prevent coagulation. Then the tube is centrifuged to isolate plasma, and calcium is added to reverse the citrate to allow coagulation. Finally, factor III (tissue factor, which promotes thrombin formation from prothrombin) is added and clot time determined optically. The international normalized ratio is the ratio of the patient's time to the laboratory standard.

Partial thromboplastin time is a measure of blood clotting in the intrinsic pathway (often triggered by trauma within the vascular system). Blood is collected in a citrate or similar tube to prevent coagulation. The tube is centrifuged, and plasma is harvested as for PT. Calcium is added to allow coagulation, again as for PT. Rather than tissue factor, the intrinsic pathway is activated using kaolin or similar. Kaolin is a clay mineral and is historically significant such that it used to be part of the name of the test.

Drug testing for particularly illicit drugs is often performed by enzyme multiplied immunoassay technology (EMIT). A blood sample is incubated briefly with antibody to the molecule being assayed. If the drug is in the sample, antibody will bind to it. Then the sample is incubated briefly with a known concentration of the drug bound to a specific enzyme. This competes with any of the drug in the patient sample for the antibody. When the antibody binds to the drug–enzyme complex, it inhibits the enzyme from working. Then, a substrate is added that is converted by enzyme not bound to antibody and produces a measurable color. If there was drug in the sample, it will have bound to the antibody so that there is more unbound enzyme to react with substrate and produce color. If there was no drug in the sample, the antibody will bind to the added drug–enzyme, which then will not react with substrate.

Therapeutic drug monitoring is increasingly performed so that effective yet not toxic doses are used. Due to specifics of patient size, volumes of distribution, and metabolism in multiple pathways, purely weight-based administration is not reliable. Analytic methods include EMIT, discussed previously, but also high-performance liquid chromatography (HPLC), fluorescence polarization immunoassay, and enzyme-linked immunosorbent assay (ELISA). Applications of mass spectrometry are also used. HPLC involves passing a sample through a column that contains material which allows separation based on specific flow rate through the column. Passage time through the column depends on interaction of the agent of interest with the column constituents.

IMMUNOLOGICAL STUDIES

Serologic testing in neurology is usually performed to identify antibodies that can be responsible for a range of diseases, including myasthenia gravis, Lambert–Eaton myasthenic syndrome, and neuromyelitis optica spectrum disorder.

The concept of autoimmune disease is a huge topic itself, but briefly, autoantibodies are made against a physiological target and that attack results in loss or disturbance of function of the target. In many circumstances, the antibody and target are known but the precise mechanism of how the attack causes the clinical disorder is not.

Measurement of the autoantibodies is performed in a number of ways depending on the purpose. For most of these, there is measurement of the levels of specific antibodies in the blood or body fluids. Among the tools used is ELISA, as discussed previously. Some of the most important antibody–disorder associations are as follows:

- Autoimmune
  - Acute disseminated encephalomyelitis (ADEM): Anti-MOG-IgG antibody seems to play a role in some cases of ADEM. The mechanism may be that an infection shares antigenic determinants with this and other antibodies, producing immune attack.
  - Multiple sclerosis (MS): Considering how common MS is, the specific mechanism of immune attack is not well understood. It is believed that antibodies affect oligodendrocytes, resulting in central demyelination.
  - Myasthenia gravis: Antibodies bind to the nicotinic acetylcholine receptor, resulting in reduced activation of the muscle with neuromuscular transmission.
- Paraneoplastic
  - Neuromyelitis optica spectrum disorder: IgG antibodies to aquaporin-4, a water channel protein, are responsible for approximately 80% of cases. The attack results in inflammation in affected regions.
  - Lambert–Eaton myasthenic syndrome: Antibodies to the voltage-gated calcium channel reduce the calcium entry, thereby reducing transmitter release.
  - Paraneoplastic cerebellar degeneration: Antibodies to Purkinje cells result in cellular death and degeneration.

DNA ANALYSES

DNA analysis is used in neurology mainly for diagnosis of inherited diseases. Because this analysis can be done at any time in the individual's life with the same results, this can be done not only in symptomatic patients but also in patients who are asymptomatic.

DNA has been studied since the late 1800s, but only since the 1980s has DNA been used to identify specific individuals. The techniques have not been flawless, with some mistakes made despite careful study, and some mistakes are due to technical sloppiness. The purpose of identification is usually forensic, determining whether a sample of fluid is from a specific individual. Although more than 99% of DNA is in common with others of our species, there are sufficient differences to be able to distinguish one person from another.

DNA analysis is performed by first extracting DNA from the sample under evaluation. There are several mechanisms to accomplish this. Then select DNA is amplified to facilitate evaluation, and this is typically done by polymerase chain reaction (PCR). The DNA is denatured into smaller strands. Then DNA primers are used to begin the replication process, with nucleotides adding to the 3' end of the chain. Serial performance of the same cycle results in geometric increases in number of copies.

For DNA fingerprinting or profiling, the sections are short tandem repeats, which are groups of small chains of three to five bases of the same nucleotide. Because this procedure is performed on DNA from different individuals, there are different repeats in different individuals. So the pattern of repeats can identify a specific individual with near perfect accuracy, as long as there are sufficient samples and no laboratory error.

Calculating differentiation probabilities can result in excessively high theoretical capability because the technique assumes no linkage, but this is not always the case. For example, identical twins would have the same DNA pattern. Brothers would have similar DNA patterns, so misinterpretations are possible. Laboratory error occurs not just from technical performance but also from sample collection.

INFECTIOUS DISEASE

Identification of infections is by smears, cultures, nucleic acid analysis, and/or serology. Neurologically, we deal with organisms that can be difficult to culture and are often not seen on a microscope slide, so they are often diagnosed by serology.

The most common test for many of the organisms we see is PCR. This is performed as described previously. However, for RNA viruses, reverse transcriptase PCR is used.

Serology is the measurement of antibodies in blood or cerebrospinal fluid (CSF), and it is especially useful when PCR is negative despite the patient being infected. PCR testing is subject to the absence of the virus in certain parts of the body. For example, the virus can be in the mouth but not in the nose. It can also give false-positive results after viral infections have resolved because bits of viral RNA or DNA may still be present in the body. For this reason, PCR tests are no longer used to determine whether it is safe for someone who has been infected with COVID-19 to return to work.

PATHOLOGY

Neuropathology has a multimodal approach to diagnosing neurologic disease antemortem and postmortem. Many of the tests described previously are used for diagnosis and are not discussed further.

Tissue obtained depends on the location and setting. Of course, amounts of tissue differ depending on biopsy versus autopsy. With a living patient, the accuracy of securing a diagnosis has to be balanced against the risk of the excision of neural tissue necessary for sensation and function.

Histology is evaluation of the tissue, typically optically. While features of specific cells are evaluated, the cellular structure is interpreted in the context of the tissue features, including vasculature, connective tissue, and other morphological factors. Special stains of antibodies linked to fluorescent molecules or enzymes can be used to target pathogens, malignant cells, or even proteins associated with diseases in order to make them more visible.

Cytology is the analysis of individual cells, typically with cellular structure assessed in the absence of supportive tissue. This can be run on body fluids, washing, or smears.

CLINICAL CORRELATIONS

Herpes simplex virus (HSV) encephalitis is suspected when a patient presents with confusion, often with fever and/or seizures. CSF analysis usually shows an increase in the number of white blood cells (pleocytosis). Often, culturing the virus from the CSF requires a significant amount of time and does not always work even in cases in which the person clearly is infected. Although brain biopsy is considered definitive for diagnosis of HSV encephalitis, this is invasive, results are not available for days, and false negatives are common. One test that can piggyback on the cytology test of the CSF is a PCR test searching for HSV RNA. The magnification ability of the PCR technique makes this a highly sensitive test. It also typically takes only 1 day to perform. Testing of the CSF using cytology and PCR is now considered to be adequate for confirmation of the diagnosis.

Toxoplasmosis is a cause of structural lesion, especially in patients with HIV. Diagnosis is often suspected when a patient presents with seizures, encephalopathy, or focal signs and has a history of HIV or is believed to be otherwise at risk. Occasional patients are seen without HIV. IgG and IgM antibodies can help determine infection and differentiate acuity of the disorder. IgM antibodies are generated acutely in the disorder, and IgG antibodies are generated later. So if the patient has IgM without IgG antibodies, then the diagnosis of acute infection is considered. Confirmation is made when the IgG antibody appears. If neither antibody appears, then infection is believed to be less likely, although repeat testing is typically performed.

Myasthenia gravis is suspected with weakness and fatigue, especially when there is ptosis or other ocular motor weakness. Because the causative agent in most patients is antibody directed against the acetyl-choline receptor or against a receptor protein muscle-specific tyrosine kinase, immunoassay for these antibodies is supportive of the clinical diagnosis, typically in conjunction with electrophysiological tests. There are seronegative cases that are diagnosed when these and other antibodies are negative yet clinical and electromyography (EMG) findings are strongly supportive of the diagnosis, especially if the EMG includes single-fiber EMG study.

Neurologic Therapeutics

EVAN M. JOHNSON AND KARL E. MISULIS ■

HISTORY AND DEVELOPMENT OF NEUROLOGIC THERAPEUTIC TECHNIQUES

Humans are infused with a drive to help, and that is likely part of the reason for our advancement as a species. This is not an inevitable part of societal development but, rather, undoubtedly had evolutionary advantage. Although we try to be scientific about our approach, even today, many of our advances are from trial-and-error approaches.

Trephination—drilling a hole in the head—may be the earliest documented neurologic technique designed to help humans. There are signs of this occurring thousands of years ago, but there is no record of the reason for the procedure. We presume it was to aid some medical condition, but that is not known with certainty.

The appearance of some skulls that have been found, with bone healing around the wounds, indicates that some of these people lived for significant lengths of time after the procedure. This bone healing usually takes weeks to months. Note that it is not easy to drill a hole in the skull. In some cultures, tools were specifically used for the purpose of elevating the scalp. In what is present-day Peru, the Nazca used a blade that looks remarkably like a similar but larger blade used for severing heads.

It is possible that the hole could be therapeutic for a patient with sub-dural or epidural hematoma, but the proportion of patients with injuries who would benefit would be low. However, if this technique is performed and the patient survives, it is an unfortunate quirk of human nature to assume that the intervention made the difference. At the time when trephination was performed, the scientific method had not yet been developed.

Almost as soon as the consistent ability to produce a shock was developed, electric shock was used on patients afflicted with a variety of diseases. In ancient Egypt, it was known that electric shocks delivered by submersion in water with a species of fish that was prevalent in the Nile River and surrounding areas produced muscle contraction. Although the use of these electric fish is documented on the walls of some of the Egyptian buildings, there is uncertainty regarding what this treatment was used for. Some historical information suggests arthritis, but use for paralysis was subsequently tried. This would seem appropriate because a paralyzed limb moves in response to the fish electricity, so perhaps that might jump-start the nervous system.

More recently, but still in old times, electric shocks were used for patients with psychiatric disorders, often by inducing seizures. The first documentation of this was in the mid-18th century, although there were isolated reports of this being used prior. Benjamin Franklin was one of multiple science-minded people of his day who used electric shocks to treat patients with paralysis, presumably from stroke, although there are no clear clinical data.

Techniques to improve clinical outcome have been tried since prehistory. Only within the past two centuries, and perhaps less, has a scientific approach been taken. It is human nature to want to help. It is human nature to try treatments even if the mechanisms are unknown; there are many treatments that have been proved to be helpful but for which the exact mechanism is not known. But it is also human nature to be anchored in our thought processes, clinging to techniques that we and our teachers and heroes have used. The techniques discussed in this chapter have been proven to be helpful by careful science. Although no study is flawless, the studies supporting the techniques discussed here are quite good.

Nonmedical treatments are used for a wide variety of disorders, and in neurology, neurostimulation techniques are prominent. They are most commonly used for epilepsy and movement disorders. The stimulation techniques are not first line for any of these conditions, but they can be an adjunct to medical management or of benefit when medical management has not helped.

The concept is that stimulation modifies the response of central pathways. Early studies showed that vagal nerve stimulation produced enhancement of metabolism in select centers of the brain, especially the thalamus. Because this is a relay station for cortical efferents and a component of many control circuits, a theory is that disordered response of neuronal circuits can be improved by this form of stimulation. Interestingly, the effective target for responsive neurostimulation is often a region in the thalamus that exhibits significant metabolic changes with vagal nerve stimulation.

Despite much research, the precise mechanism of action of all of these neurostimulation procedures remains elusive. Yet the responses seen are so marked that we are confident that we are on the right track but do not fully understand the physiology.

DEEP BRAIN STIMULATION

What Is Deep Brain Stimulation?

Deep brain stimulation (DBS) is a method of targeted neurostimulation via electrical impulse generation. It is carried out as a neurosurgical procedure, in which a thin wire filament is placed through the skull cavity that terminates at the desired location for implementation. It can be thought of as a targeted version of electroconvulsive therapy (ECT) with permanent lead implants allowing for a relatively small electrical field to be generated in an area of interest corresponding to the activity desired.

The precise mechanism of DBS is not fully understood. By generating a small targeted electrical field, it is thought that local neuronal activity is modulated, which may include inhibition or increased firing potential at the site of interest.

DBS offers an alternative to surgical resection or ablation of targets as well as the less directed ECT.

How Did Deep Brain Stimulation Come About?

The premise of directed brain activity modulation dates back several centuries, as inducing seizures (convulsive therapy) was an early attempt at therapy for various psychiatric disorders. Medicinal in origin, this gave way to ECT in the early 20th century. More focused delivery of electrical stimulation in the 1900s would bring about the advent of modern DBS before the end of the century.

Preceding achievements helped pave the way. The development and refinement of the stereotactic frame in the late 1940s championed by Ernest A. Spiegel and Henry T. Wycis allowed for precise preplanned minimally invasive neurosurgical lead placements with a controlled trajectory and depth. The development of the implantable cardiac pacemaker in 1958 was a second key innovation because its basic premise of an implantable battery with leads extending into a target of interest for focused electrical stimulation would lay the groundwork for DBS technology.

The initial application of the technology was primarily for pain management and psychiatric disorders. The resultant effects from targeting specific nuclei and from the use of particular frequency ranges were slowly realized over this period of time. The use was at times diagnostic rather than therapeutic, with investigators hoping to identify sites for subsequent ablation procedures. Over time, it was realized that chronic continuous stimulation may represent a form of long-term therapy for various conditions.

DBS was tested with external power sources in the 1970s ahead of implantable power generator systems in the 1980s. DBS had previously shown effectiveness in the treatment of pain, but studies in its use for Parkinson disease (PD) had been underwhelming. In 1987, French neurosurgeon and physicist Alim-Louis Benabid demonstrated that direct stimulation of the thalamus at higher frequency (100 Hz) was effective in reducing tremor in the PD patient he was operating on. This discovery quickly brought about a new wave of interest and work, ultimately leading to U.S. Food and Drug Administration (FDA) approval of DBS surgery for essential tremor in 1997 and for PD in 2002.

What Sets Deep Brain Stimulation Apart from Other Therapeutic Modalities?

Unlike most therapeutic procedures, DBS is an adjustable localized stimulation-specific subcortical brain tissue treatment, giving the provider added precision, control, and a degree of reversibility. Ablation treatments permanently destroy an area of tissue, and adverse effects may be permanent. Extracranial treatments may have limited penetration or effective depth. DBS permits access to deeper structures such as the basal ganglia and thalamus, and the programmable stimulation allows the provider to adjust settings to provide the patient optimal relief of symptoms with minimization of side effects.

How Does Deep Brain Stimulation Work?

In DBS, a thin electric lead is placed through a coin-sized burr hole in the skull at a particular trajectory through the cranium. Its tip is implanted within a particular site of interest at a desired angle to best fit the expected electrical field to the desired tissue without overlapping neighboring brain structures. The site of interest is chosen based on the patient's condition, desired symptom control, and known comorbidities. The ventral intermediate nucleus of the thalamus (VIM) is typically targeted for essential tremor. The subthalamic nucleus (STN) and globus pallidus internus (GPi) are the most common sites targeted for PD. In obsessive–compulsive disorder (OCD), certain nuclei within the thalamus or GPi may be appropriate targets.

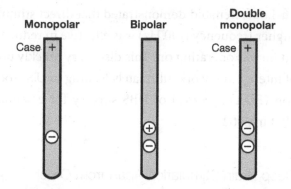

Figure 14.1 DBS configurations. A small sampling of various possible stimulation configurations in DBS; only active leads are shown. (*Left*) Monopolar, in which a single contact is negative (site of electric stimulation delivery) and the case is positive (for the impulse generator battery); (*middle*) bipolar, in which one contact is negative and another positive; and (*right*) double monopolar, in which two contact points are set as negative, creating a more oblong shape of stimulation. The repelling force of the positive contacts shapes the electrical stimulation into a narrower disc.

The field size and shape are primarily controlled by contact configuration and delivered current. The shape of the field charge may be influenced by the manner in which the contacts are employed. Configurations of a selection of DBS leads are shown in Figure 14.1.

Use of a single contact is a unipolar configuration that creates a roughly spherical field. Double monopolar configuration makes use of two contacts and creates a more oblong field. Bipolar configuration employs both anode and cathode contacts, which can produce a more disc-shaped field restricting vertical spread. Directional leads, as mentioned previously, act as a "searchlight," "steering" the stimulation outward in a specified direction and minimally in an undesired one. An example of this is the use of a lateral segmented contact in a GPi target to minimize spread, and thus side effects, medially to the internal capsule. Interleaving involves rapidly alternating unipolar fields at different contact levels.

The stimulation pattern may also be customized for maximum desired response with minimal adverse effects. The conventional pattern used in DBS is an asymmetric biphasic waveform in which a short-duration cathode pulse is followed by a brief pause and then a longer duration anode pulse. Numerous waveforms may also be utilized, including a symmetric biphasic waveform in which the anode and cathode pulse are of the same duration, a symmetric waveform without a pause between the anode and cathode pulses, or waveforms in which the cathode pulse precedes the anode.

Stimulation may be composed of pulses occurring at a frequency with fixed intervals, such as 130 Hz or approximately one pulse every 7.7 msec, or, less commonly, nonregular random intervals or a burst pattern composed of several bursts at high frequency with short intervals followed by a longer interval between pulse bursts.

Pulse width refers to the duration of each moment of stimulation. Indirectly, pulse width is considered in practice to deliver the stimulus to a larger field of tissue.

Systems may deliver stimulus based on programmed constant voltage versus constant current settings. Constant voltage is the more traditional method, in which the generated current is determined by impedance, as with Ohm's law. In DBS, resistance comes not only from the hardware but also from the biological tissue it encounters. Swelling (edema) in the vicinity of the lead contacts can lower impedance and permit electric stimulation to spread over a greater volume and a higher current. Fibrous scar tissue can increase impedance. Usually, impedance in DBS systems is approximately 1,000 ohms, but it can be less than 700 ohms or more than 5,000 ohms in certain situations, depending on both biology and hardware. In a constant current system, that can lead to large changes in delivered current. Constant current, by contrast, uses feedback to adjust battery voltage to maintain a consistent programmed current (mA), mitigating changes in impedance. This provides a more consistent level of stimulation for the patient, but it can invite unexpected drainage of the battery.

Typical DBS parameter settings for movement disorders range from 1 to 4 V (constant voltage) versus 0.5 to 5 mA (constant current), 60–450 us pulse width, and 60–185 Hz frequency.

Total electrical energy delivered (TEED) can be an important equation for clinicians to be aware of, especially because it can help anticipate battery drain in systems. TEED is calculated as voltage squared times pulse width times frequency divided by impedance:

$$\text{TEED} = V^2 \times \text{PW} \times \frac{f}{Z}$$

where V is the voltage, PW is pulse width, F is frequency, and Z is impedance.

In a monopolar review, a programmer may stepwise assess each contact at various settings (normally current at intervals with a set frequency and pulse width) to gain a sense for which contact may offer the widest therapeutic window: beneficial effect at lower current and absence of adverse effects unless at a higher current. Once this preferred contact is determined, directional leads can be reviewed to further optimize the therapeutic window, ideally providing a good response with minimal settings. If the desired response is not realized, pulse width and frequency may also be adjusted to find a "sweet spot." There are ongoing studies and academic debate regarding the effects of adjusting these settings.

What Are the Components of a Deep Brain Stimulation System?

The basic components of the DBS system include a battery-powered pulse generator, an electrode lead implanted within the brain, and an insulated wire connecting the two.

The implanted pulse generator (IPG) is typically surgically implanted beneath the skin of the chest. Modern IPGs offer programmability after implantation with accompanying near-field wireless communication devices. Modern devices are capable of powering two separate leads, allowing for bilateral DBS with a single IPG.

The connecting wire runs beneath the skin from the head to the side of the IPG.

The lead is surgically implanted from the skull to the area of interest. Conventional DBS leads are roughly 1.3 mm in diameter, composed of platinum–iridium wires with nickel alloy connectors, and coated in polyurethane. The choice of these materials maximizes conduction properties while minimizing heavy metal toxicity. Each lead typically contains four contact nodes toward the terminus in the innermost location of the brain where the target is located, referred to as a quadripolar configuration. Standard leads have 1.5-mm contacts that are spaced 0.5– 1.5 mm apart from each other. Relatively recently, leads have become more complex segmented contacts that allow more controlled electric field shaping; these are known as directional leads. Another development, multiple independent current control, employs dedicated current sources for each contact node, increasing the precision that may be used in programming.

Once implanted, the properties of the applied stimulation can be modified through programming to maximize benefits while minimizing adverse effects. The major targets in DBS (VIM, STN, and GPi) lie close to neighboring structures, so a stimulation field of effect can produce unwanted effects if it overlaps with these nearby tracts. For example, inadvertent stimulation of the internal capsule may produce paresthesias or involuntary muscle contraction, and a field of effect involving the optic tracts may affect vision.

What Are Some of the Common Uses of Deep Brain Stimulation in Neurology?

- PD
- Essential tremor
- Epilepsy
- Dystonia
- OCD
- Pain

DBS is currently being reviewed for use in depression, dementia, addiction, and other psychiatric conditions such as schizophrenia.

DBS can offer significant symptom control in tremor-predominant idiopathic PD patients who benefit from dopamine replacement therapy. It can improve rigidity and tremor in these patients via stimulation at the STN. Patients with significant issues with balance, ambulation, mood disorder, or cognitive decline may experience better overall results with GPi. Dystonic patients may experience some relief of symptoms with GPi. Essential tremor patients and even some patients with refractory parkinsonian tremors may be best served with a VIM target.

DBS targets in epilepsy may vary and are specific to the patient. Centromedian thalamic nucleus is a common target in generalized epilepsy, anterior nucleus of the thalamus can be common for partial or secondary epilepsy, and the hippocampus is a common target for mesial temporal lobe epilepsy.

Surgery

The DBS procedure is a multistep process. It begins with clinical evaluation of the patient, determining potential candidacy and potential site targeting. When a candidate is chosen and a site determined, magnetic resonance imaging (MRI) is frequently the next step to allow visualization and localization of the target nucleus for presurgical planning. In many instances, fiducial markers are placed such that the surgeons can establish three-dimensional localization and accurately and precisely plan for the site of entry, trajectory, and depth to place the electric lead at the nucleus of interest. Specific MRI sequences allow for optimal viewing of target site anatomical boundaries, which is crucial for presurgical planning. The structures adjacent to the target site may vary in tissue property, making it difficult to clearly delineate all borders with a single MRI sequence. Advanced techniques continue to be developed to help produce ideal presurgical images. Although 1.5 and 3 Tesla magnets are more commonly seen in current clinical MRI scanners, powerful 7 Tesla magnet MRIs offer ultra-high-field imaging that permits visualization of intrathalamic nuclei, such as the VIM. Susceptibility mapping can take advantage of tissue iron content and permit enhanced imaging of the STN. A fast gray matter acquisition T_1 inversion recovery sequence provides contrast between the GPi and surrounding subcortical structures.

On the day of the surgery, a small hole is drilled into the skull at the point of entry, and the lead is placed at a specific trajectory and depth that correlates with the nucleus of interest. Intraoperative imaging may take place to confirm appropriate lead location. Depending on site and surgeon, DBS may be implanted either as an asleep or awake procedure. In an asleep procedure, the patient is anesthetized throughout, and lead placement is confirmed via intraoperative imaging and/or microelectrode recording. Microelectrode recording is a technique in which local neuronal firing patterns are assessed at various depths. Described firing patterns allow the proceduralists to confirm a desired location along the tract. In an awake procedure, the patient is awake for the procedure, and after lead placement, the DBS is tested by asking the patient to perform tasks so that effective outcome can be confirmed.

The IPG is commonly implanted under the skin of the patient's chest, with the leads run under the scalp to it. This can be done as part of the lead implantation procedure or as a separate procedure later. The IPG is a complex piece of hardware. It is the power source of the system, can be a rechargeable battery in newer models, and is programmable. A patient or provider wireless interface device communicates with the IPG and allows the provider to program the settings of the DBS unit: contact configuration, voltage, current, pulse width, frequency, and more.

Postoperatively, the DBS is not turned on until a subsequent initial programming visit. This interval allows for the patient to recover from the surgical procedure and for postsurgical changes to settle, as these can confound programming. The patient is then seen in clinic, and optimal contact settings are determined to maximize benefit without overstimulation.

Adverse Effects

As with all procedures that penetrate the skin, there are risks of infection and bleeding, including intraparenchymal hemorrhage, with surgical lead placement. In some patients, postoperative cerebral edema or lesion can lead to new seizure activity. Common postoperative effects may include speech, cognitive, or gait difficulties in some patient.

If surrounding brain tissue is affected by the generated impulse inadvertently, adverse effects can be seen. Bilateral stimulation may also lead to these issues.

Stimulation beyond the desired nucleus involving adjacent structures may lead to various effects reflective of the structures affected. Paresthesia, muscle pulling, vision impairment, significant emotional change, speech impairment, and other effects could occur.

The human body's reaction to the foreign DBS lead material can affect the electrode–tissue interface and efficiency of the stimulation. Glial cells are seen in the general inflammatory reaction, altering the tissue environment of the target site.

Like other implantable electrical devices, the use of DBS in MRI can pose a safety concern for patients. Powerful magnetic fields applied to a conducting wire induce an electrical current. The generation of this current can lead to heating of the metal and subsequently cause damage to surrounding tissue. The DBS lead tips are a particular area of concern for this. In addition, application of the MRI magnetic fields can cause dysfunction and damage to the IPG if operating. Most modern devices permit safe MRI with appropriate settings applied.

Future Developments

Future developments in DBS are a topic of excitement for both patients and clinicians. One development being tested currently is closed-loop feedback control, allowing for real-time programming adjustments to maximize benefit and reduce battery usage. In PD, there has been observation in some patients of a distinct pattern of neuronal firing in specific areas of the brain that correlates with "off" periods, or times in which patients' symptoms are inadequately controlled by medicine or stimulation. There is ongoing research in developing DBS leads that are able to continuously monitor for these characteristic local field potentials using contact segments not delivering stimulation. The detection of the characteristic firing pattern would then prompt a preprogrammed adjustment in delivered stimulation, with the intention to provide more help when needed.

The development of devices with a miniaturized IPG allowing for limited intracranial placement without extension to a separate unit placed in the chest is also expected to occur in the future.

VAGAL NERVE STIMULATION

Vagal nerve stimulation (VNS) is performed by placing stimulating wires around the left vagus nerve and an implanted stimulator. The parameters are set by a wireless controller and adjusted according to beneficial and adverse effects.

The stimulator usually is set to cycle with 30 seconds on time and 5 minutes off time. Timing parameters and intensity are adjusted for best effect.

In newer versions of VNS, stimulation is induced by accelerated heart rate, considering that this can be an indicator of seizure.

In addition to epilepsy, VNS is used for refractory major depressive disorder.

RESPONSIVE NEUROSTIMULATION

Responsive neurostimulation (RNS) is used for some epilepsy patients who have not responded well to medical management. RNS relies on recorded rhythms of the brain to guide generation of brief bursts of electrical stimulation in response to patterns that the controller identifies as epileptogenic. This is an effective modality resulting in marked seizure reduction. However, there is a learning curve for the device, so configuration and function are maximally effective only over time. Use is in patients with medically refractory epilepsy, and it is not first-line treatment.

RNS has been considered for other disorders, including pain and some psychiatric disorders, but this is not indicated use and only time and scientific progress will determine if any of these other indications are appropriate.

TRANSCRANIAL MAGNETIC STIMULATION

What Is Transcranial Magnetic Stimulation?

Transcranial magnetic stimulation (TMS) is the application of a strong focused magnetic field over a specific area of a patient's scalp to stimulate a targeted area of the brain to promote a desired long-term effect. It is a therapeutic procedure most commonly used to alleviate psychiatric disorders.

How Did Transcranial Magnetic Stimulation Come About?

Transcranial magnetic stimulation can be thought of as a gentle descendant of ECT. Two major principles of these treatments were built upon findings from more than 200 years ago. In the late 1700s, Luigi Galvani helped establish that biological tissue responded to electricity. In a particularly cruel study that would not be seen in school science classes today, he tied frogs to a metal fence and observed that lightning strikes triggered leg contractions. Alessandro Volta, well known for his studies in electricity, would further this work and its indication that animals generated an internal form of electricity for communication between organs via the nervous system.

The other major principle was Michael Faraday's groundbreaking work establishing the relationship between magnetism and electricity. The concept of electromagnetic conduction and induction revolutionized science. Among his many works, the Faraday disk showed that electricity could be produced by rotation of a magnetized disk, and Faraday's iron-ring coil showed that electric current could induce a magnetic field that could in turn induce an electric current in a nearby circuit. This particular proof of mutual induction is the underlying basis of TMS.

As discussed in the section on ECT, direct electric stimulation predominated medical research and therapy throughout most of the 1900s. This work helped identify specific functional centers of the brain, such as the homunculus map of the motor strip. ECT was applied for numerous neurological and psychiatric disorders with early success, but it became socially stigmatized and began to fall out of favor by the 1970s. Although it would rebound, this fall from grace at least partially helped usher in TMS as an alternative.

The lion's share of credit for the advent of TMS has historically been given to Anthony Baker. He helped develop a system that utilized a magnetic field pulse intended to induce an electric current within the brain, rather than the application of a direct electrical stimulus. In 1985, he demonstrating the ability to painlessly stimulate targeted areas of the motor strip in a human patient. In subsequent years, this work continued to be validated and developed, and it led to therapeutic applications. With the backing of multiple studies during the 1990s, TMS became an approved therapy for medically refractory depression in the United States in 2008.

How Does Transcranial Magnetic Stimulation Work?

The simplest idea behind TMS is that a high-energy electrical pulse is used to create a strong but focused magnetic pulse near a specific area of a patient's scalp in order to induce an electrical stimulus in a desired area of the brain. It is believed that a series of sessions, or repetitive TMS, promotes network changes within the brain that will alleviate mood disorders such as depression.

Transformer

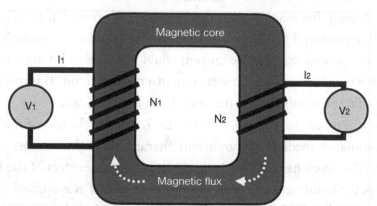

Figure 14.2 Transformer coil. Electrical transformers receive electrical current from one power source at a specific voltage and can either increase or decrease that voltage in an output electrical current. The magnetic flux through the center of the transformer is determined by the number of winds or turns on the input source, and a different number of turns on the output end will result in a different voltage.

To go into more detail, TMS utilizes a high-voltage system to generate a 2 or 3 Tesla magnetic field at a coil by the patient's scalp. To generate the powerful electrical current within the electromagnetic coil, a TMS system requires a step-up transformer, as shown in Figure 14.2.

Transformers attached to neighborhood power lines are step-down, reducing the AC delivered across power lines to a safer 120 V for residential use. Transformers are designed using Faraday principles, where multiple coils are wound around a magnetic core. Faraday's law of induction states that the voltage of an electromagnetic coil is related to the number of turns or winds of the electrical coil. Therefore, within a transformer, a system with one coil with a certain number of turns coupled to a magnet and a second coil with a different number of turns can induce an electric current with a different voltage. Depending on how it is built, the transformer can receive electricity from a power source and then provide an output of either an increased (step-up) or decreased (step-down) voltage.

The system then employs a capacitor to store the required electrical energy so that it is ready to be delivered at the push of a button. Capacitors are at their heart a dam within a river of electrical current. A noncon-ducting block is placed within a circuit, where the electric current is still driven by voltage to continue but is unable to cross. Capacitors are built so as to store a specific amount of energy in a system, allowing engineers to prepare a system to deliver that specific amount of electricity to an appli-cation such as is the case in TMS.

A high-voltage cable connects the power source to the TMS coil, typ-ically a 6- or 7-cm-diameter "butterfly" or "figure eight" coil that is rem-iniscent in appearance to an old-fashioned key to a wind-up toy. When activated, the capacitor releases the high-voltage current to the TMS coil as a brief 250-μsec pulse, creating a 2 or 3 Tesla magnetic pulse that is directed perpendicular to the coil and toward the underlying brain. This reminds of the "right hand rule" in physics, in which one can re-member the induced magnetic field direction by curling the fingers of one's right hand. When the curled fingers follow the path of the electrical current within the coil (thumbs-up if counterclockwise, thumbs-down if clockwise), the thumb points in the same direction as the magnetic field vector.

As the TMS electric current dissipates, the produced magnetic pulse will then induce a current in the nearby secondary circuit: the patient's cortex. The neurons affected by the TMS are limited to the cortex within 5 cm of the coil. The electric current produced is far less than what the TMS power source delivered, making this entire process not unlike a step-down transformer. This follows logically because biological brain tissue is a fairly inefficient electrical circuit relative to the TMS power supply and TMS coils.

What Are the Applications of Transcranial Magnetic Stimulation?

TMS is currently FDA approved for medically refractory depression, delivered as a series of treatment sessions over several weeks. An advantage seen over ECT is that TMS appears to spare patients of the memory loss that is often seen in those undergoing ECT.

TMS was also approved for the treatment of medically refractory Migraines in 2013; and, more recently, for obsessive-compulsive disorder (OCD) in 2018.

The use of TMS is being explored for additional psychiatric conditions, including bipolar disorder, schizophrenia, and PTSD. It is also being investigated for potential use in numerous neurological disorders including dementia, stroke, ALS, multiple sclerosis, and epilepsy. A potential limiting factor in the application of TMS as opposed to ECT is the more restricted depth of effect.

Adverse Effects

Transcranial magnetic stimulation has generally been found to be a well-tolerated procedure in the majority of patients. Adverse effects are considered to be relatively rare but have been reported to include seizures, syncope, pain, impaired cognition, and hypomanic states. The applied magnetic field of TMS can interact with implanted devices, including DBS circuits and pacemakers, and its use is discouraged in these patients.

ELECTROCONVULSIVE THERAPY

What Is Electroconvulsive Therapy?

Electroconvulsive therapy is a procedure in which electrical current is applied to a patient's cranium in order to induce seizures in a controlled environment. The procedure can be unilateral or bilateral. Typical therapy is composed of a series of procedures performed over the course of several weeks. ECT is most often utilized for refractory psychiatric conditions, such as severe depression or schizophrenia, but its use has been studied for several neurological conditions as well.

How Did Electroconvulsive Therapy Come About?

Electroconvulsive therapy originated as a novel form of convulsive therapy in the first half of the 20th century. Effective medical intervention for psychiatric conditions has been challenging for centuries, including today.

The idea of convulsive therapy was rooted in observations dating at least as far back as the Roman Empire that there was a residual effect on the psyche following a seizure. This led to trials aimed at inducing a seizure with the intent to alleviate conditions such as pain, psychosis, or depression. Numerous methods were tried, including camphor as early as the 16th century, insulin shock therapy, pentylenetetrazol, and other chemicals—collectively referred to as pharmacological shock therapy. These treatments reached peak use in the early to mid-1900s before falling out of favor to ECT.

Applying electrical shocks as a form of therapy was experimented with without significant success following the introduction of electricity in society. ECT was developed in the late 1930s by Italian professor Ugo Cerletti, who had already been involved in pharmacological convulsive therapy research. Cerletti was inspired by the use of powerful electrical shocks used on livestock in slaughterhouses to induce a seizure and post-ictal state to ease the process of killing the animal. He and his assistant, Lucio Bini, at Sapienza University of Rome experimented with electrically induced seizures on animal subjects and then applied this technique to human subjects as a new form of convulsive therapy. Their work demonstrated significant benefit on certain psychiatric conditions following a series of 10–20 ECT sessions, and this treatment was approved for use in the United States and Europe in 1940. ECT became an attractive alternative to pharmacological shock therapy given reduced cost and the perception of increased convenience and safety.

Improvements to the technique followed, which included the use of temporary muscle relaxants to reduce the incidence of injury, specifically fractures; sedatives to reduce anxiety; replacement of continuous current delivery with a brief pulse for increased control delivery of shock; and the development of unilateral electrode placement. ECT often results in retrograde amnesia, and patients often do not recall the experience of the procedure.

Negative portrayals in media led to stigma and reduced use of ECT by the mid-1970s. The American Psychiatric Association issued task force recommendations in 1978, 1990, and 2001 guiding appropriate patient selection, consent, training, and technique implementation of ECT for specified conditions. Its efficacy for medically refractory conditions has led to continued investigation for its application to other difficult-to-treat diseases.

What Sets Electroconvulsive Therapy Apart from Other Therapeutic Modalities?

Electroconvulsive therapy is a non-invasive, nondestructive technique of neuromodulation, making it an attractive alternative to irreversible procedures such as lobotomy.

How Does Electroconvulsive Therapy Work?

Electroconvulsive therapy is carried out by placing two electrodes on a patient's head. Unilateral ECT involves both electrodes on the same side of the head, and bilateral ECT commonly has the electrodes placed at either temple. Unilateral ECT is believed to have fewer adverse effects, although it is also less effective.

Electroencephalography (EEG) and electrocardiology (ECG) leads are additionally placed to permit intraprocedural monitoring. The patient is given a short-acting sedative and paralytic as an additional safety precaution.

The specific current and voltage applied are most often individually determined for each patient based on observation of that patient's seizure threshold. This is so the stimulation delivered is not excessive with increasing adverse effects but, rather, sufficient to reliably induce a seizure. Standard practice is to apply an electrical stimulus 1.5 times the determined seizure threshold in bilateral ECT and 12 times the threshold found in unilateral ECT.

Typically, the electrical stimulus used in ECT is approximately 800 mA and supplies up to several hundred watts.

Modern shock voltage is given for a duration of 0.5 msec, whereas conventional brief pulse is 1.5 msec.

As a point of comparison, it is helpful to revisit the basics of electricity. It is ultimately the power delivered that should be compared, which is calculated as the product of voltage and current. Voltage is the magnitude of the attractive force compelling electrons to travel from one location to another. Current quantifies the flow of electrons. An example of high voltage but low current is static shock: The voltage can exceed 20 kV, but the current is so low due to high resistance that the transfer is harmless (low power). Traditional disposable batteries (AA, AAA, C, and D) are 1.5 V each, and they summate if they are combined in series, such as in a flashlight. You can hold both ends safely because the voltage is so low and the resistance of human skin is so high (>1,000 ohms) that very little current (1.5 mA in this example) or power (0.02 W) can be produced. By contrast, a bolt of lightning is approximately 300 million V and approximately 30,000 A, producing 9 GW. In comparison, household current is 120 V and 15 A, producing 1,800 W.

Last, the total energy delivered depends on how long a wattage is applied. This is why an ultra-brief lightning bolt is potentially survivable and why a household electrical outlet can potentially result in a fatal electrocution. This final premise is why ECT applies an electrical current for only a brief period of time (0.5–1.5 msec).

What Are the Components of an Electroconvulsive Therapy System?

An ECT system consists of electrodes, lead wires, and an interface system enabling control of the delivery settings. A transformer and capacitor combine to supply the controlled delivery of electrical stimulus. The transformer applies Faraday electromagnetism principles, discussed in further depth in the section on TMS, to adjust the voltage of the power source to a desired level. The capacitor builds up and stores a desired level of charge that can subsequently be delivered when ready.

What Are Some of the Common Uses of Electroconvulsive Therapy in Neurology?

The most established use of ECT is in severe psychiatric conditions that have not responded to other forms of treatment, such as therapy or medication. ECT is FDA approved for major depressive disorder, catatonia, bipolar disorder and mania, schizophrenia, and schizoaffective disorder. It has also been used in neuroleptic malignant syndrome.

Research studies have indicated benefits from ECT in treatment-refractory OCD and the motor symptoms of PD and related conditions, including Lewy body dementia and progressive supranuclear palsy. ECT has also been considered a potential salvage therapy for certain cases of intractable epilepsy.

Adverse Effects

The most apparent adverse effect of ECT doubles as its primary intended effect: inducing seizure. The potential for physical injury prompted modern-day safety measures, including intraprocedural monitoring with EEG and pre-procedural administration of short-acting paralytics and sedatives. Atropine may also be given to reduce salivation because patients may not be able to protect their airway and risk aspiration. It is not uncommon for patients to report muscle soreness following the procedure.

ECT has a reported mortality rate of 6 per 1,000 treatments, and cardiac dysfunction is associated in approximately one-third of patients. Arrhythmia and asystole are specific cardiac events observed. Although not a strict contraindication, the presence of cardiac comorbidities is weighed in patient selection. Intraprocedural ECG monitoring is an important measure included for this reason.

Memory loss and impaired cognition are frequently reported following procedures, both typically temporary in nature. Anterograde amnesia, the ability to create and retain new memories, has been described in the time period closely following the procedure before resolving. Retrograde amnesia, the loss of previous memories, has been described to involve the memories in the months leading up to the procedure in some patients.

NEUROLOGIC REVASCULARIZATION

Stroke treatment has advanced in recent years, and what was routine treatment in the recent past is far out of date. The ability to intervene and improve outcome with stroke has resulted in development of stroke centers that parallel trauma centers in organized approaches to stroke care. Like the trauma centers, this has resulted in increasing centralization. We argue that one of the major technical advances in neurologic care is this organization. Now, with acute stroke as well as trauma, a high proportion of patients are cared for at facilities that see sufficient volumes and have organized care teams so that outcomes can be better even if a particular patient is not a candidate for reperfusion therapy.

Thrombolytic Therapy

In the early days of attempts at stroke intervention, heparin was given routinely to patients with ischemic stroke, with the logic that it would reduce thrombus formation and perhaps facilitate breakup and dissolution of the thrombus that was present. Subsequent study showed that this was not supported by data and should be abandoned except for a few uncommon exceptions, such as venous sinus thrombosis.

For ischemic stroke, recombinant tissue plasminogen activator (TPA or alteplase) was shown to be of real benefit in controlled studies. Initially, the maximum time interval from onset of symptoms to administration of TPA was 3 hours, and currently it is 4½ hours. This is routinely timed from when the patient was last known well, meaning observed directly to have no new neurologic deficits. This did not have to be a medical evaluation; the patient's report or that of family or another who was in the presence of the patient would suffice. If TPA was given within those parameters, then it could be of marked benefit. If, on the other hand, it was given far after the time when the patient was normal, the benefit was less and the risk was increased. Therefore, most stroke centers have a hard line on this time frame, except for select patients who have stroke noted on wake-up and the last time of normalcy is not known. If the non-contrasted CT head is negative for early infarct, edema, or blood, then MRI can be performed STAT, and findings on that may suggest onset within 4½ hours and the patient may be considered for TPA. This is not standard at all facilities; it is so mainly at select stroke centers.

TPA works by binding to fibrin in a thrombus and converting plasminogen to plasmin. Plasmin is the major enzyme responsible for breakdown of thrombus. Plasminogen is the pro-enzyme and is itself inactive. Facilitating this transformation results in increased breakdown of thrombus.

With more data, the use of TPA is expanding, and there are more agents being developed that can facilitate thrombus dissolution.

Endovascular Therapy

Endovascular therapy (EVT) for acute ischemic stroke is performed when there is impairment in one or more arteries with thrombus formation. The thrombus may be occlusive (no flow distal to it) or non-occlusive (in which at least some flow bypasses the thrombus). EVT is performed by a trans-arterial approach, usually from radial or femoral artery. There are a number of devices. Some of them are retrievers, designed to capture a clot and have it removed. Stents are also used when there is stenosis but no clot that can be retrieved.

STEREOTACTIC RADIOSURGERY

Stereotactic radiosurgery (SRS) is a form of ablation therapy that utilizes focused radiation to destroy a tissue target in such as a way as to minimize damage to the surrounding body as much as possible.

The radiation can be delivered in several forms, such as an array of focused gamma rays (Gamma Knife), high-energy monochromatic X-ray beams produced from a linear accelerator, or proton beams.

SRS using a gamma beam array employs approximately 200 beam trajectories that converge on the tissue of interest so that any one specific beam delivers a tolerable radiation dose along its path, but the combined dose at the target tissue converging point is sufficient to kill the cells there. Careful planning is undertaken to achieve this, using a presurgical high-resolution image such as MRI and advanced computer modeling, often Monte Carlo simulation. Computer simulation models estimate absorbed dose across each specific particle track-length segment, calculating total dose imparted within the volume and targeted tissue. Once the therapy plan is created and verified, the patient undergoes an outpatient procedure. The radiosurgery system looks not unlike a CT or MRI, but the patient bed only allows for the head to enter the bore and is met by a "helmet." This head frame acts as a collimator with a large number of tunneled beam paths. The gamma radiation is delivered along the numerous beam paths in either one or multiple sessions. This method allows for the most precise delivery of ablative energy for small targets. Example applications of this include ablation of small tumors, including metastatic disease or epileptic foci. This procedure can permit targeted destruction of pathologic brain structures so small or so multiple that conventional surgical resections may not be feasible.

Linear accelerators (LINACs) are an older and more established form of radiotherapy. A diagram of a LINAC is shown in Figure 14.3. In this, X-ray beams are generated from very high-energy electron beams. This is not dissimilar to the previous discussion on X-ray beam generation for radiography and CT, but with an important difference in desired ultimate beam energies. Whereas diagnostic systems optimally utilize the lowest

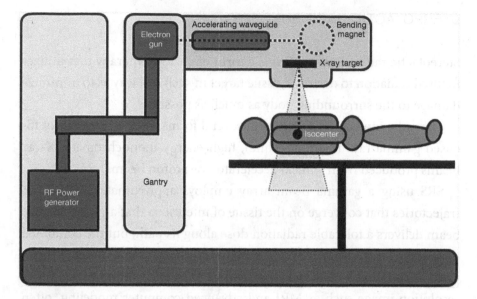

Figure 14.3 Linear accelerator. A linear accelerator generates a high-energy electron beam that is ultimately delivered to a target such as tungsten to create high-energy X-rays, typically for radiation therapy. The electrons are accelerated through a vacuum tube across a series of channels with alternating current to continuously provide a repulsive electric force behind and attractive electric force in front of the electrons. The electrons are subject to additional energy via microwave oscillation from a magnetron, adding to the kinetic energy of the particles.

ionizing radiation sufficient to penetrate body tissue for detection and image reconstruction, radiotherapy is designed to deliver therapeutic energy levels sufficient to destroy pathologic tissue such as tumors. These systems also produce an electron and X-ray beam with a much narrower energy range (monochromatic). To achieve this boosted electron beam acceleration, pulsed microwave energy is produced by a Magnetron and delivered to increase their kinetic energy. A Magnetron may sound like the leader of enemy robots from a children's cartoon, but it is actually a high-power vacuum tube that generates microwave radiation from the interaction of electrons with a magnetic field within a series of cavities that help create a self-oscillating environment. The electrons within the LINAC are propelled toward the target anode (typically tungsten) via a series of alternating polarity tubes, such that they continuously have a forward pulling attraction and antegrade repelling force. The resultant electron beam energies vary based on system and design, but they often range from 4 to 25 MeV. The subsequent X-ray beams are thus in the mega electron volt range—considerably higher energy than diagnostic imaging beams (usually 120 keV or less). Similar to gamma-based SRS, several shaped beam path trajectories are modeled and used to take advantage of convergence to spare normal nontargeted tissue as much as reasonably achievable. A rotating gantry is typically employed with an LINAC and X-ray tube.

Proton beam therapy is less common in medical centers due to its cost and need for specialized centers. The principles behind proton beam acceleration are similar to those for electrons, although with oppositely charged particles. The other key is the large difference in mass between the two particles. A proton mass is more than 1,800 times that of an electron. This means that at comparable velocity, the momentum of an accelerated proton dwarfs that of an electron. In practice, directed proton beams have increased penetrance and can deliver far more energy to a tissue target than can electrons. Although proton beams cannot penetrate tissue as fully as X- or gamma rays, this means that they do not continue to deliver damaging energy beyond the target of interest, sparing healthy tissue. Proton therapy is considered a strong therapeutic tool for ablating pathologic tissue such as tumors within sensitive organs, including the brain, eye, or lung.

FOCUSED ULTRASOUND

Focused ultrasound is another form of ablation procedure. As the name implies, high-intensity ultrasound waves are focused on a pathologic tissue of interest in order to deliver sufficient energy to heat and destroy the cells. Diagnostic sonography can identify the area of interest, or different modalities such as MRI may be used. To heat the tissue, the ultrasound delivers lower energy continuous sound waves rather than the conventional pulsed waves in imaging. This form of therapeutic ablation dates back many decades, and it informed later techniques. Many of the subcortical targets of modern DBS were originally ablation targets. In addition, focused ultrasound remains a viable option for disabled patients who are not otherwise considered to be candidates for DBS surgery. Another well-known use of focused ultrasound is lithotripsy, in which the focused high-energy sound waves are used to break up obstructive renal stones too large to pass naturally.

Chapter 2

1. There are also digital circuits, in which the same data flow is accomplished, but in this circumstance the flow is performed by gates, where data of a specific digital character triggers a function to be executed which produces a digital output. This concept is used in computer science, but for our purposes, it is not important for analog technology.
2. Those well-versed in hydrodynamics may object to this analogy because when the pressure is high enough, flow becomes no longer laminar, becoming turbulent, therefore increasing the resistance again, but this is an illustrative analogy and not meant to be an exact model.
3. Black body radiation is electromagnetic radiation emitted by black or opaque structures, and we cannot see the radiation as light yet can otherwise perceive warmth. This is because the radiation is infrared, so our eyes, with quite narrow spectrum of photon wavelength perception, cannot see it.

Chapter 3

1. Sung SF, Hung LC, Hu YH. Developing a stroke alert trigger for clinical decision support at emergency triage using machine learning. *Int J Med Inform.* 2021; 152: 104505.
2. Powers WJ, Rabinstein AA, Ackerson T, et al. Guidelines for the early management of patients with acute ischemic stroke: 2019 update to the 2018 Guidelines for the Early Management of Acute Ischemic Stroke: A Guideline for Healthcare Professionals from the American Heart Association/American Stroke Association. *Stroke.* 2019; 50(12): e344–e418.
3. Rae-Grant A, Day GS, Marrie RA, et al. Practice guideline recommendations summary: Disease-modifying therapies for adults with multiple sclerosis: Report of the Guideline Development, Dissemination, and Implementation Subcommittee of the American Academy of Neurology. *Neurology.* 2018; 90(17): 777–788.
4. Kevin Johnson [additional credits].

5.   Okazaki EM, Yao R, Sirven JI, et al. Usage of EpiFinder clinical decision support in the assessment of epilepsy. *Epilepsy Behav*. 2018; 82: 140–143.
6.   Makin JG, Moses DA, Chang EF. Machine translation of cortical activity to text with an encoder–decoder framework. *Nat Neurosci*. 2020; 23(4): 575–582.
7.   Sirsat MS, Fermé E, Câmara J. Machine learning for brain stroke: A review. *J Stroke Cerebrovasc Dis*. 2020; 29(10): 105162.

CHAPTER 4

1.   Michelson DJ, Ashwal S. The pathophysiology of stroke in mitochondrial disorders. *Mitochondrion*. 2004; 4(5–6): 665–674. doi:10.1016/j.mito.2004.07.019
2.   Terasaki Y, Liu Y, Hayakawa K, et al. Mechanisms of neurovascular dysfunction in acute ischemic brain. *Curr Med Chem*. 2014; 21(18): 2035–2042.

CHAPTER 5

1.   Janik EL, Jensen MB. Every man his own electric physician: T. Gale and the history of do-it-yourself neurology. *J Neurol Res Ther*. 2016; 1(2): 17–22. doi:10.14302/issn.2470-5020.jnrt-15-910
2.   Finger S. Benjamin Franklin, electricity, and the palsies: On the 300th anniversary of his birth. *Neurology*. 2006; 66(10): 1559–1563. doi:10.1212/01.wnl.0000216159.60623.2b
3.   Silverman MG, Scirica BM. Cardiac arrest and therapeutic hypothermia. *Trends Cardiovasc Med*. 2016; 26(4): 337–344.
4.   Hypothermia after Cardiac Arrest Study Group. Mild therapeutic hypothermia to improve the neurologic outcome after cardiac arrest. *N Engl J Med*. 2002; 346(8): 549–556. doi:10.1056/NEJMoa012689. Erratum in: *N Engl J Med*. 2002; 346(22): 1756.
5.   Bernard SA, Gray TW, Buist MD, et al. Treatment of comatose survivors of out-of-hospital cardiac arrest with induced hypothermia. *N Engl J Med*. 2002; 346(8): 557–563. doi:10.1056/NEJMoa003289
6.   Nielsen N, Wetterslev J, Cronberg T, et al.; TTM Trial Investigators. Targeted temperature management at 33°C versus 36°C after cardiac arrest. *N Engl J Med*. 2013; 369(23): 2197–2206. doi:10.1056/NEJMoa1310519
7.   Dankiewicz J, Cronberg T, Lilja G, et al; TTM2 Trial Investigators. Hypothermia versus normothermia after out-of-hospital cardiac arrest. *N Engl J Med*. 2021; 384(24): 2283–2294.15

Action potential—change in potential due to a cascade of channel openings and closing overshooting zero potential.

Aliasing—error in representation of a signal with an improper sampling interval.

Amplifier—circuit element that uses a power source to increase the energy of a signal.

Analog-to-digital converter—circuit element that samples analog signals and converts them into digital equivalent.

Aneurysm—dilation of blood vessel due to weakness of the wall.

Angiogram—imaging study of blood vessels.

Anion—ion with negative charge.

Artifact—signal contamination by potentials not from the intended generator.

Artifact rejection—capacity of devices to suppress potentials not part of the response of the system being recorded.

Artificial intelligence—capacity of computer systems to perform tasks that would usually be attributed to humans.

Atom—smallest particle of an element.

Atomic mass—number of protons plus neutrons in an atom.

Atomic number—number of protons in an atomic nucleus; determines which element it is.

Auditory evoked potential—study that examines the pathways of acoustic stimuli from the acoustic nerve through the brainstem.

Averaging—function of devices to average the Y values of a series of trials to reach a most likely value.

Axon—output protrusion of a neuron that synapses on other cells.

Bond—connection between elements; can be covalent or ionic.

Capacitance—ability of a circuit element or a segment of excitable tissue to separate change.

Capacitor—circuit element that stores energy by separation of charge.

Cation—atom with a positive charge.

Charge—ability of an element or device to respond to an electric field.

Circuit—network of elements that provide for movement of charge.

Clinical decision support—protocol in electronic health records that provides assistance to providers.

Computer—device that performs mathematical and logical operations.

Conductor—material that has atomic properties which are conducive to electron movement and hence conduct electric current.

Contrast agent—substance used to highlight or otherwise differentiate certain pathologies on imaging studies.

Detector—device that transduces a signal into electrical data for analysis.

Diode—circuit element that allows for current to flow in only one direction.

Dual energy—for computed tomography scan, use of electron beams with two different energy levels to help differentiate composition of certain relatively radiopaque signals.

Electrode—connector between a biologic system and a device.

Electroencephalography—study that examines electrical activity from the brain.

Electromyography—study that examines physiologic function of the neuromuscular system.

Electron—subatomic particle with a charge of –1.

Electron beam—stream of electrons typically produced by the cathode.

Electronic health record—application suite for health care that usually includes multiple access methods to multiple databases to give a record across an enterprise or health care system.

Equilibrium potential—potential at which the electrical and chemical gradients balance and there is no net charge movement of the index ion(s).

Evoked potentials—studies that examine the response to the nervous system to stimuli, usually sensory (somatosensory, auditory, and visual) but sometimes motor.

Far field—potential that is recorded beyond proximity to the generator, in this sense. In physics, far field (and near field) have somewhat different definitions.

Filter—circuit element or computer algorithm that suppresses potentials which are not part of the signal of interest.

Fluoroscopy—video imaging of the body using X-rays.

Gray matter—cerebral tissue with rich composition of cell bodies.

Hydrodynamics—study of fluid movements.

Impedance—resistance due to resistance of elements plus reactance of changing current flow.

Inductor—circuit element that stores energy in the form of a magnetic field.

Intracranial pressure—level of fluid pressure in the brain.

Isotope—versions of an element that have different numbers of neutrons, although they have the same number of protons.

Machine learning—use of computers to solve problems without being specifically programmed to identify the solution.

Magnetic resonance imaging—medical imaging using magnetic fields and radio waves.

Myelin—insulating sheaths of axons in both the central and the peripheral nervous system.

Myelogram—imaging of the spinal column involving injection of contrast agent into the spinal fluid.

Near field—in recording, potentials produced immediately adjacent or very close to the electrode.

Nerve conduction—study in which stimulation and recording are able to examine motor and sensory nerve fiber action potential propagation.

Neurotransmitter—chemical released at synaptic junctions intended to stimulate or inhibit electrical activity in the target neuron.

Neutron—subatomic particle with neutral charge.

Noise—in electrical terms, a seemingly random series of signals that interfere with identification of the signal of interest.

Nonconductor—material that is not conducive to electron movement and cannot conduct electricity.

Nuclear magnetic resonance—property of nuclei to produce an electromagnetic signal in response to a magnetic field.

Ohm—unit of resistance.

Positron—antiparticle of electron, same mass but positive charge.

Power spectrum—distribution of the power (energy) of a spectrum of frequencies.

Proton—subatomic particle with a positive charge.

Resistor—circuit element that dissipates energy usually in the form of heat.

Semiconductor—material that conducts better than nonconductor but less well than conductor.

Signal-to-noise—ratio of the amplitude of the signal of interest to background electrical noise.

Somatosensory evoked potential—study that examines responses of the spinal and brain neurons to electrical stimulation of peripheral nerves.

Stimulator—electronic device used to activate nerves, usually for nerve conduction or evoked potential studies.

Stroke—damage to the brain or spinal cord or eye from damage to arterial or venous structures supplying the area.

Synapse—connection between two neurons that facilitates signal transmission.

Synaptic transmission—chemical transmission from one nerve to another cell.

Tomography—mapping usually part of the body as if a slice or cross section.

Transistor—circuit element with gating properties; used for multiple purposes.

Ultrasound—sound higher than humans can hear; used as a stimulus to visualize especially body soft tissues.

Vasospasm—constriction of blood vessels, which can reduce blood flow.

Venogram—image of the veins usually of the brain, usually using computed tomography or magnetic resonance imaging.

Visual evoked potential—evoked potential elicited by visual stimuli, usually to assess the anterior visual pathways.

Voltage—a measure of the electromotive force or a difference in potential between regions.

Volume conduction—transmission of biologic signals through tissue before reaching the sensors.

Watershed—ischemia usually of the brain due to lowered perfusion pressure rather than acute occlusion of a blood vessel.

White matter—brain matter that is myelinated.

X-ray—form of electromagnetic radiation with an energy higher than ultraviolet light but less than gamma rays.

For the benefit of digital users, indexed terms that span two pages (e.g., 52–53) may, on occasion, appear on only one of those pages.
Figures are indicated by an italic *f* following the page number.